P9-BVM-862

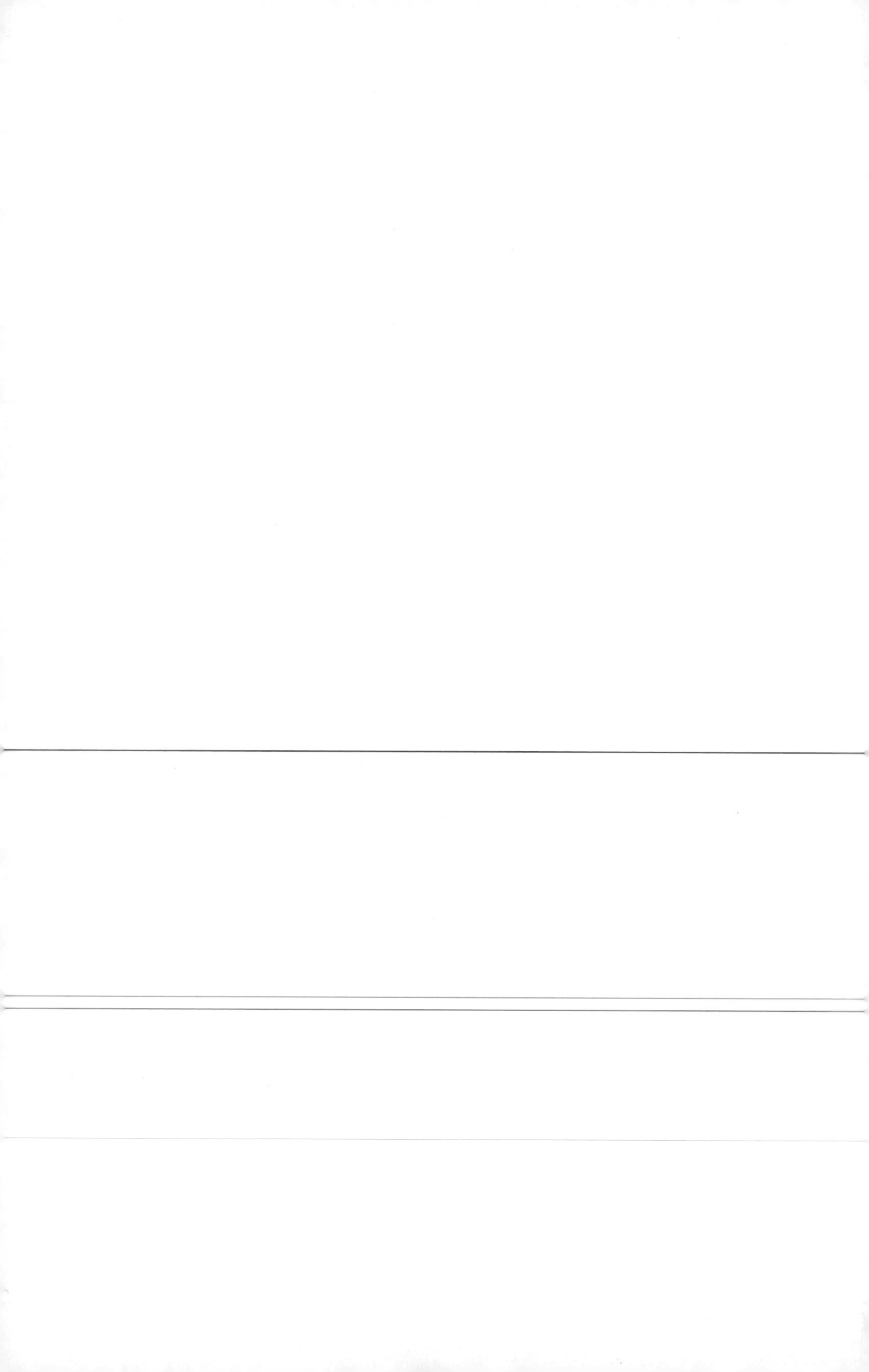

MOLECULAR BIOLOGY
INTELLIGENCE
UNIT

Discovering Biomolecular Mechanisms with Computational Biology

Frank Eisenhaber, M.D., Ph.D.
Bioinformatics Group
Institute of Molecular Pathology
Vienna, Austria

LANDES BIOSCIENCE / EUREKAH.COM
GEORGETOWN, TEXAS
U.S.A.

SPRINGER SCIENCE+BUSINESS MEDIA
NEW YORK, NEW YORK
U.S.A.

Discovering Biomolecular Mechanisms with Computational Biology

Molecular Biology Intelligence Unit

Landes Bioscience / Eurekah.com
Springer Science+Business Media, LLC

ISBN: 0-387-34527-2 Printed on acid-free paper.

While the authors, editors and publisher believe that drug selection and dosage and the specifications and usage of equipment and devices, as set forth in this book, are in accord with current recommendations and practice at the time of publication, they make no warranty, expressed or implied, with respect to material described in this book. In view of the ongoing research, equipment development, changes in governmental regulations and the rapid accumulation of information relating to the biomedical sciences, the reader is urged to carefully review and evaluate the information provided herein.

Springer Science+Business Media, LLC, 233 Spring Street, New York, New York 10013, U.S.A.
http://www.springer.com

Please address all inquiries to the Publishers:
Landes Bioscience / Eurekah.com, 810 South Church Street, Georgetown, Texas 78626, U.S.A.
Phone: 512/ 863 7762; FAX: 512/ 863 0081
http://www.eurekah.com
http://www.landesbioscience.com

Printed in the United States of America.

9 8 7 6 5 4 3 2 1

Library of Congress Cataloging-in-Publication Data

Discovering biomolecular mechanisms with computational biology / [edited by] Frank Eisenhaber.
p. ; cm. -- (Molecular biology intelligence unit)
Includes bibliographical references and index.
ISBN 0-387-34527-2 (alk. paper)
1. Biomolecules--Structure. 2. Bioinformatics. 3. Molecular biology. I. Eisenhaber, Frank. II. Series: Molecular biology intelligence unit (Unnumbered)
[DNLM: 1. Computational Biology--methods. 2. Genomics--methods. 3. Molecular Biology--methods.
QU 26.5 D611 2006]
QP517.M3D57 2006
572--dc22

2006012698

Dedicated to Natalya Georgievna Esipova
and Vladimir Gayevich Tumanyan
from the Engelhardt Institute of Molecular Biology in Moscow.

CONTENTS

Section II: Complementing Biomolecular Sequence Analysis with Text Mining in Scientific Articles

Section III: Mechanistic Predictions from the Analysis of Biomolecular Networks

EDITOR

Frank Eisenhaber
Bioinformatics Group
Institute of Molecular Pathology
Vienna, Austria
Introduction, Chapter 3

CONTRIBUTORS

Miguel A. Andrade
Ottawa Health Research Institute
Ottawa, Canada
and
European Molecular Biology Laboratory
Heidelberg, Germany
and
Department of Bioinformatics
Max Delbrück Center
for Molecular Medicine
Berlin-Buch, Germany
Chapter 5

Saurabh Asthana
Genetics Division
Department of Medicine
Brigham and Women's Hospital
and Harvard Medical School
Boston, Massachussetts, U.S.A.
Chapter 10

Vladimir B. Bajic
Knowledge Extraction Lab
Institute for Infocomm Research
Singapore
Chapter 4

Peer Bork
European Molecular Biology Laboratory
Heidelberg, Germany
and
Department of Bioinformatics
Max Delbrück Center
for Molecular Medicine
Berlin-Buch, Germany
Chapter 5

Harmen J. Bussemaker
Department of Biological Sciences
and
Center for Computational Biology
and Bioinformatics
Columbia University
New York, New York, U.S.A.
Chapter 6

Liran Carmel
National Center for Biotechnology
Information
National Library of Medicine
National Institutes of Health
Bethesda, Maryland, U.S.A.
Chapter 11

Vidhu Choudhary
Knowledge Extraction Lab
Institute for Infocomm Research
Singapore
Chapter 4

Weng Keong Choy
Department of Biological Sciences
National University of Singapore
Singapore
Chapter 4

Thomas Dandekar
Abteilung Bioinformatik
Biozentrum, Am Hubland
Universitaet Wuerzburg
Wuerzburg, Germany
Chapter 2

Toni Gabaldón
Nijmegen Center for Molecular Life Sciences
Center for Molecular and Biomolecular Informatics
Nijmegen, Netherlands
Chapter 1

Martijn A. Huynen
Nijmegen Center for Molecular Life Sciences
Center for Molecular and Biomolecular Informatics
Nijmegen, Netherlands
Chapter 1

Rajaraman Kanagasabai
Knowledge Extraction Lab
Institute for Infocomm Research
Singapore
Chapter 4

Edda Klipp
Max Planck Institute for Molecular Genetics
Berlin Center for Genome-Based Bioinformatics
Berlin, Germany
Chapter 7

Eugene V. Koonin
National Center for Biotechnology Information
National Library of Medicine
National Institutes of Health
Bethesda, Maryland, U.S.A.
Chapter 11

S.P.T. Krishnan
Knowledge Extraction Lab
Institute for Infocomm Research
Singapore
Chapter 4

Marko Marhl
University of Maribor
Department of Physics
Maribor, Slovenia
Chapter 7

Archana Meka
Department of Biological Sciences
National University of Singapore
Singapore
Chapter 4

Bijayalaxmi Mohanty
Knowledge Extraction Lab
Institute for Infocomm Research
Singapore
Chapter 4

Hong Pan
Knowledge Extraction Lab
Institute for Infocomm Research
Singapore
Chapter 4

Carolina Perez-Iratxeta
Ottawa Health Research Institute
Ottawa, Canada
and
European Molecular Biology Laboratory
Heidelberg, Germany
and
Department of Bioinformatics
Max Delbrück Center for Molecular Medicine
Berlin-Buch, Germany
Chapter 5

Karin Schleinkofer
Abteilung Bioinformatik
Biozentrum, Am Hubland
Universitaet Wuerzburg
Wuerzburg, Germany
Chapter 2

Christian Schlötterer
Institut für Tierzucht und Genetik
Veterinärmedizinische Universität Wien
Austria, Europe
Chapter 9

Stefan Schuster
Friedrich-Schiller University Jena
Department of Biology
and Pharmaceutics
Section of Bioinformatics
Jena, Germany
Chapter 7

Berend Snel
Nijmegen Center for Molecular
Life Sciences
Center for Molecular
and Biomolecular Informatics
Nijmegen, Netherlands
Chapter 1

Shamil Sunyaev
Genetics Division
Department of Medicine
Brigham and Women's Hospital
and Harvard Medical School
Boston, Massachussetts, U.S.A.
Chapter 10

Sanjay Swarup
Department of Biological Sciences
National University of Singapore
Singapore
Chapter 4

Sin Lam Tan
Knowledge Extraction Lab
Institute for Infocomm Research
Singapore
Chapter 4

Edward N. Trifonov
Genome Diversity Center
Institute of Evolution
University of Haifa
Chapter 8

Pardha Sarathi Veladandi
Department of Biological Sciences
National University of Singapore
Singapore
Chapter 4

Yuri I. Wolf
National Center for Biotechnology
Information
National Library of Medicine
National Institutes of Health
Bethesda, Maryland, U.S.A.
Chapter 11

Zhuo Zhang
Knowledge Extraction Lab
Institute for Infocomm Research
Singapore
Chapter 4

Li Zuo
Knowledge Extraction Lab
Institute for Infocomm Research
Singapore
Chapter 4

INTRODUCTION

Bioinformatics:

Mystery, Astrology or Service Technology?

Frank Eisenhaber*

Abstract

Mathematical interpretation and integration of experimental data for the goal of biological theory development has had little, if no, impact on previous progress in life sciences compared with the sophistication of experimental approaches themselves. The genesis of recent spectacular breakthroughs in molecular biology that led to the discovery of the enzymatic function of several nonmetabolic enzymes illustrates that this relationship is beginning to change.

The development of high-throughput technologies, for example of complete genome sequencing, leads to large amounts of quantified data on biological systems without direct link to biological function that require formalized and complex mathematical approaches for their interpretation. The research success in life sciences depends increasingly on the ability of researchers in experimental and theoretical biology to jointly focus on important questions. Currently, theoretical methods have best chances to contribute to new biological insight independently of experiments in the area of genome text interpretation and especially for gene function prediction. Experimental studies can help progress in the development of theoretical methods by providing verified, sufficiently large and variable sequence datasets for the exploration of sequence-function relationships.

Introduction

To caricature, the typical research process in life sciences consists of periodic repetitions of weeks/months of bench work by a PhD or postdoctoral student followed by an hour of looking at the results by the lab head after which the coworker again disappears into the cold room or behind the microscope with new directives. Generations of life scientists have been educated that the most important goal consists in producing "hard", quantitative experimental data describing biological structures and processes. Pure theoretical efforts directed at biological data analysis are believed to add little more than intellectual speculation or a colorful illustration in the form of a graph or an alignment. The biological theory itself has remained logically simple and with little or no mathematics or formal structure. Typically, all creativity has been directed into sophistication and rationalization of experimental procedures and techniques for the wet lab. This type of life science has successfully produced breakthroughs and will, apparently, continue to stay the major source of new biological insight in the near future.

*Frank Eisenhaber—Research Institute of Molecular Pathology, Dr. Bohr-Gasse 7, A-1030 Vienna, Republic Austria. Email: Frank.Eisenhaber@imp.univie.ac.at

Discovering Biomolecular Mechanisms with Computational Biology, edited by Frank Eisenhaber.

This situation is especially astonishing for people that come from more formalized sciences such as physics where a typical experiment is preceded by months of calculations and computer simulations. Such research is necessary in this area to derive the most interesting research targets and to check the consistency of new hypotheses with existing knowledge. There had been several waves of efforts to inject mathematics into life sciences, for example statistics (beginning with Mendel's ratios), kinetics (of enzymes and ligand binding, of transport systems, in population dynamics) or 3D biomolecular structure modeling (together with quantum chemistry, QSAR studies and molecular dynamics; especially in context with the hypothesis of DNA double-helical structure). Although each of these waves have enriched life sciences in aspects, neither one has become a continuous source of qualitatively new biological knowledge or has made biology a truly theoretical, a quantitative and predictive science.

Beginning with the 1960s, yet another stream of efforts focused on the esoteric topic of analysis of text strings representing the monomer sequences of proteins and nucleic acids and of the evolution of these strings after multiple single-point mutations.[1,2] Thanks to these pioneering efforts, theoretical concepts, computational methods and sequence databases have been established that allow the prediction of function for experimentally uncharacterized genes from their sequence, most importantly, together with the quantification of the prediction error (prediction reliability) in probabilistic terms.[3] The impact of this development is perceived in different ways by various parts of the generally wet lab-focused life science community depending on personal background and experience: as *mystery*, *astrology* or *service technology*. None of these three ways is a really appropriate assessment for the recent step in the difficult development of life sciences towards a formalized theory of living systems as the discussion below will attempt to show.

Mystery

Sometimes, success stories are sensed euphorically as a *mystery* by those scientists that receive a tremendous boost in their experimental work from a function prediction. At the background of general weakness of theory in life sciences, it is indeed perceived as a bolt from the blue by the experimental life science research community that a number of recent scientific breakthroughs in biology have originated from theoretical studies for gene function prediction. Several instances of discoveries of enzyme activities for a number nonmetabolic proteins, typically without any previous hint or suspicion from experimental findings, are remarkable evidence for the growing predictive power of theoretical biology.

Important science-organizational and cognitive aspects of this process towards new biological knowledge can be illuminated by viewing some recent examples of enzymatic function assignment to nonmetabolic enzymes. Three stories with considerable biological impact, namely

1. the discovery of the molecular function of Fringe in Notch signaling,
2. the determination of the protease domain of separin triggering the transition from metaphase to anaphase during the cell cycle, and
3. the understanding of heterochromatin formation as initiated by the histone methyltransferase activity of the Su(var)3-9 homologues, are described in brief in Boxes 1, 2 and 3.

There are more of such nontrivial findings, and it is not possible to give a complete list here. For example, a C-terminal domain in yeast protein dot1p was assigned to the SAM sequence family with suggested methyltransferase activity. The loss-of-function phenotype of the dot1 gene (disruption of telomere silencing) implied a possible role in the posttranslational modification of histones.[4] Indeed, a biochemical assay was able to show that dot1p does methylate histone H3 at Lys79.[5] As the authors acknowledge, the previously published theoretical report was critical for their decision to launch the experimental test.

In another case, the yeast protein eco1p was found critical for the establishment of cohesion between sister chromatids,[6] but the biological experiments did not give any hint with respect to a possible molecular function of eco1p. Sequence analysis studies pointed to an

Box 1. The Fringe Story

As early as January 1997, Peer Bork and three of his collegues[27] reported that a sequence domain in the *D. melanogaster* Fringe and Brainiac proteins as well as in their vertebrate homologues might have glycosyltransferase activity. Relying on simplified but statistically rigorous models of molecular sequence evolution implemented in the BLAST[3] and MoST[28] programs, they could show that there is negligibly small probability of error if the corresponding sequence domains of Fringe are considered members of the same protein family together with a number of bacterial monosaccharide transferases including, for example, Lex1 from *H. influenzae*.

Whereas the sequentially close neighbors have been collected with BLAST (a pairwise sequence comparison tool), the link between the eukaryan and bacterial subfamilies has been quantitatively assessed with MoST (an early sequence profile technique). The collection of families of homologous sequence segments was much more difficult, labour intensive, and less straightforward in summer 1996 than with today's more automated alignment and profile generation tools such as PSI-BLAST.[13,29] Based on statistical measures of sequence similarity and taking into account the conserved sequence pattern of hydrophobic/hydrophilic residues and secondary structure predictions, Peer Bork and colleagues finally concluded that all proteins involved share certainly the same type of 3D structural fold. But whether the same type of catalytically active center is attached to this scaffold remained hypothetical although the conservation of sequence motifs with functional residues and the molecular biological context suggested this as highly probable.[27]

This finding was completely surprising for the field. The Notch gene was one of the first genes to be identified in *D. melanogaster* being discovered by T.H. Morgan and colleagues in 1916. Its role in the spatial control of tissue-patterning events and in developmental signaling through the cytolemma has long been known. More recently, the secreted protein Fringe[30,31] was shown to differentially modulate Notch sensitivity towards the ligands Delta and Serrate/Jagged.[32-34] Although the phenomenon was well described at the phenotypic and cellular level, nothing was understood with respect to molecular mechanisms. We do not know about the scales of hypothesis-driven research generated by the prediction of Fringe's catalytic activity but, only three years after the paper of Bork et al, the authors of two reports[35,36] admitted that they had been inspired by the theoretical finding and presented convincing evidence that Fringe does indeed change the glycosylation status of Notch during the latter's passage through the Golgi apparatus.

acetyl-coenzyme A binding site in the C-terminal domain of eco1p. It was this finding that changed the priorities in the wet lab, and the subsequent experiments showed that eco1p indeed has acetyltransferase activity and, apparently, is part of a yet unknown mitotic pathway that involves acetylation of some cohesion complex proteins.[7]

What can be learned from these stories?

i. A hypothesis derived from an extensive theoretical analysis suggested a previously unknown direction of thought and creatively enriched biological theory, not only in details but, for the given field of research, in a principal aspect. This happens everyday in formalized sciences, but it is new in biology.

ii. There is an increasing need for an altogether time- and resource-consuming effort of theoretical analysis of biological experimental data accompanying life science research projects from the outset. If completed successfully, it can lead to nontrivially new, creative directions of experimentation or directly to new biological insight. The new role of theory represents a qualitative change resulting from quantitative progress in the experimental method accumulated over decades. Large amounts of quantitative data without direct link to known biological processes at different levels of organizational hierarchy are now produced with high-throughput techniques and require a nontrivial effort for interpretation.

Box 2. The Protease for Cohesins

Cellular and nuclear divisions (mitosis) have been known since the early days of studying biological tissue with microscopes. The arrangement of sister chromosomes in the metaphase plate is shown in every cytology textbook. Nevertheless, we have learned about the molecular "glue" holding sister chromatids together and about the molecular release mechanism at the onset of anaphase only during the last few years.[37] Using yeast genetic and molecular biological methods as well as cellular microscopy, the "glue" was shown to represent a complex of four types of proteins (scc1p, scc3p, smc1p, smc3p) which dissociates from sister chromatids under the influence of a protein named esp1p at the onset of anaphase.[6,38] Even more, scc1p was shown to be proteolytically digested in an esp1p-dependent manner in an in vitro assay for reconstituting the chromatid dissociation reaction using cell extracts.[39] The nature of the protease involved remained obscure and no obvious experimental approach for finding it was in sight.

The alignment of C-terminal sequence segments of several esp1p homologues was analyzed and the strict conservation of two potentially functional residues, a histidine and a cysteine (H1505 and C1531 in esp1 respectively), was observed throughout species. Together with considerations based on secondary structure predictions and applying encyclopedic knowledge of protease families, Eugene Koonin suggested but cautiously that the C-terminus of esp1p itself might belong to a new group of proteases which is distantly related to other members the CD clan of cysteine endopeptidases with the same type of catalytic dyad.[40,41] Indeed, subsequent experiments for esp1p enzyme inhibition with site-directed mutations and with specifically designed peptide drugs demonstrated convincingly that it is the esp1p protease activity resulting in the cohesin cleavage[42] at the onset of anaphase.

iii. Not incidentally, it is the area of gene function prediction where, for the first time, theoretical data analysis has become a widely indispensable activity for planning experimental strategies affecting increasingly research efficiency in evermore more branches of life sciences. Sequencing of biomolecules has been the first high-throughput technology in life sciences. During the last decade of the 20th century, electronic sequence databases coupled with scientific literature sources have reached such a dimension and representation for different biological subfields that the theoretical meta-analysis of this data without any additional experimentation can lead to surprising new biological insight.

iv. This bioinformatics activity is the truly integrating factor for the life sciences as whole. Different subfields interact with each other via their entries in databases which are analyzed as one body of data in bioinformatics research efforts. For example, mammalian signaling was connected with bacterial biochemistry in the Fringe story and the functional characterization of the SET domain became possible with information on plant enzymes.

v. In many instances, theoretical analysis ended up in weak hypotheses that they could not be published directly due to lack of intrinsic logical rigor in their derivation. Therefore, many of such hints will never appear in computer-generated public sequence database annotations since a considerable theoretical and/or experimental effort is required to judge the significance of a below-threshold hit. At the same time, pure experimental approaches would have hardly uncovered the catalytically active domains in the near future. The true partnership between both approaches (and people from different groups and complementing backgrounds) was the major precondition of success. However, strong hierarchies and star-centered organizations typical for many life science research units will have difficulties accommodating such an interdisciplinary research style.

vi. Efforts in applied mathematical and biophysical research for biological sequence analysis which have been smiled at in the biological community have matured and, often, reliable predictions or, at least, useful hints for further experimental studies can be made. The most

Box 3. The SET Domain Function

After more than 5 years of studying the function of mice and human suv39h1 (and of the later identified testis-specific suv39h2), the vertebrate homologues of the *D. melanogaster* Su(var)3-9 proteins, with genetic and cellular biological methods, it was possible to describe in great detail their phenotypic function (a role in heterochromatin formation and stabilization) at the level of the whole organism and in each of the tissues.[43] Additionally, all genes involved have been cloned and single- and double-knockout mice have been produced.[44,45] Nevertheless, the SET sequence domain contained in all suv-proteins resisted any attempts for molecular functional characterization.

After collecting the sequence family of SET domain homologues with a variety of profile techniques, more than 140 members including six plant sequences were found. At about this time, PSI-BLAST[13] became a routine tool in sequence family collection. This search could be carried out in a simpler manner by starting a PSI-BLAST run, for example, with the *S. pombe* sequence SPAC3c7.09. Surprisingly, one of those proteins found (in *P. sativum*) was experimentally characterized as rubisco methyltransferase.[46] Unfortunately, all six plant sequences were closely similar to each other except for their N-terminus. Thus, the catalytic activity could be represented likewise by the SET domain, by a long insert dividing the plant SET domain into two parts, by the C-terminal domain, or even by a combination of those. In any of the latter three cases, we would have learned little new about suv-protein function.

Since experimental considerations did not provide more attractive alternatives, in vitro assays with suv-proteins, labelled S-adenosyl-methionine and potential substrates were set up. The positive outcome of the first tests agreeing with the hypothesis of mouse suv39h1 methylating histone H3 marked the beginning of a race of discoveries: Experimental evidence for the role of posttranslational methylation of histone tails in heterochromatin formation has been obtained.[47] This event triggered the functional characterization of the Chromodomain known to coexist with the SET domain in many proteins as binding domain. There is now convincing evidence that the chromodomain segment of the mammalian heterochromatin binding protein (HP1) binds to methylated histone H3 with high affinity and targets HP1 to nucleosomes with Lys9-methylated H3 tails.[48,49] Now, histone tail methylation has become a still expanding field of research with more than 1700 papers in PUBMED by beginning of 2005.

important hallmark (but not the only one) is represented by the technique of sequence alignment and the evaluation of mutual substitution rates of amino acids during evolution in form of substitution matrices. This advance allowed the quantification of sequence similarity in rigorous probabilistic terms, essentially a distance measure between sequences.[3] The concept of protein homology with the postulate of a common evolutionary ancestor for a family with sequentially similar gene segments is the basis for structure and function prediction by annotation transfer to uncharacterized proteins from experimentally analyzed homologues.[8,9]

The feeling of the mysterious appearance of the biological insights results from the cultural collision of the world of experimental life scientists not trained in thinking with statistical or physical categories because of the previous lack of necessity. The back story is the expectation of repeated amazing predictions with new sequence targets, an expectation that can hardly be satisfied in all instances. There are a large number of full sequences and sequence segments (as a rule of thumb, one third of an eukaryote proteome) where currently computational biology has to lay down its arms. The situation is even grimmer if the whole genome instead of the protein-coding part (only ca. 1.5% of the genome) is considered.

Self-critically, the bioinformatics community has to acknowledge that, especially in early work, the way that the theoretical conclusion was achieved was described superficially and, for those who are not in the know it can be difficult to repeat the deduction and to evaluate the judgments. The incomplete description of procedures is a general problem in life sciences. Repetition of published experimental recipes typically ends in failure. Sadly, the traditional methods section in scientific articles was moved from the place following the introduction to the end of the text and typed in small font signaling its relative insignificance.

Astrology

Multiple views of the same topic have a right for existence. Since biological sequence analysis is not a fundamental science in the same sense as, for example, physics, it is sometimes jokingly called "sequence astrology". This has several reasons:

Although Anfinsen[10] has shown with renaturation experiments that the information contained in the protein sequence is generally sufficient to determine the protein structure, attempts to derive native structures ab initio (with fundamental physical principles) from protein sequences alone without additional high-resolution experimental information (macromolecular X-ray crystallography and NMR) have practically failed due to the system's complexity. If the assumption basis is less restrictive and includes atomic force fields empirically fitted to observation from small molecules, the protein-modeling problem is more tractable but the intramolecular energy criteria become too inaccurate to distinguish between native and nonnative structures. Thus, ab initio approaches trying to compute a protein's structure directly from its sequence with fundamental physical principles are of little help in structure and, even more, in function predictions for real life examples.

Thus, computational sequence analyses in biological application studies rely on empirically established sequence-structure and sequence-function relationships:

i. The possibility of extrapolation of such correlation to uncharacterized sequences cannot fully be proven in principle, but it is tacitly assumed in practical applications. Considerable published research is devoted to establish statistical, physical and/or biological criteria to assess the reliability and limits of such extrapolations, (for example see refs. 11-14).
ii. Further, dramatic simplifications are often necessary to treat the problems rigorously with mathematical means. For example, the sequence evolution models underlying current sequence comparison and alignment techniques do, as a rule, ignore the generally weak but possibly existing mutual dependence among sequence positions.
iii. At the beginning of working on a new prediction algorithm, the researcher compiles a so-called learning set of sequences with experimentally verified structural/functional properties. Usually, he encounters the first problem here: Since only the first discovery is granted with a high-impact publication, the experimental community pays little attention to increasing the list of verified examples and the resulting learning set is (a) small and, probably, (b) does not represent the full variety of naturally occurring sequences with the same feature.[15-19] To conclude, even if the researcher honestly tries not to fool himself and his colleagues in the field with prediction rate cosmetics, the estimates of accuracy have a tendency to be too good looking, both with respect to false positive and false negative predictions.

This causes a perception problem of sequence analysis-based predictions: Experimental researchers seeing all the advertised 'well-working' prediction techniques around get disappointed since, for their specific target, the straightforwardly applied methods (especially if used with a restrictive parameter setting) appear not applicable or, in the other extreme, seem to produce obviously false output; thus, the seed of interest in nonexperimental approaches is nipped in the bud.

The typical experimental researcher also has difficulties distinguishing valuable suggestions from false hints. Here, we compare two examples of false predictions to make a frequently important point: The protein kinase activity suggested for scc2p[20] might be argued with the incomplete conservation of a short sequence motif typical for kinases but can be ruled out by structural considerations showing scc2p being a HEAT-repeat protein.[21,22] Generally, conserved short motifs involving only a handful of residues, especially with incomplete conservation of functionally important residues, are not indicative of evolutionary retained function if there is no additional argument. Thus, a kinase assay for scc2p appears not indicated.

In contrast, the suspected ras-binding activity of myr-5 is based on the significant sequence similarity over a specific ca. 100 AA-long domain to proteins known from experimental reports to interact with ras in that region.[23] Indeed, the structural similarity was further substantiated but the binding activity could not be.[24] Nevertheless, the extrapolation of function from family members leads to a reasonable, productive hypothesis since no contrary arguments such as nonconservation of known functional motifs or additional experimental data could be found.

For some people having a strong background in mathematical and physical sciences, all this combination of formally not proven sequence-function correlation, of function extrapolation and of intuition for still correct borderline cases in sequence studies seems intellectually unpleasing. Nevertheless, relatively simple sequence pattern recognition techniques produce surprisingly many reliable predictions. In this context, it appears necessary to emphasize the importance of evolutionary history in biology. As a result, the search space of real sequence alternatives appears dramatically limited compared to the theoretically imaginable set of variants compatible with fundamental laws.

It should also be noted here that experiments cannot be considered the final word in every instance either.[25] Iyer et al[26] list half a dozen cases where the molecular function predicted using well tested, statistically sound computational means is in direct contradiction to experimental results arguing for another enzymatic function (typically originally predicted from short motifs or with a threading approach). Whereas execution of an enzyme activity by nonhomologous proteins implies an (possibly unlikely) unusual chemistry, over-interpretation of experimental data or lack of additional controls might have led to the contradictory conclusion.

Service Technology

This is the third point of view with respect to bioinformatics. It is typically followed by computer science-driven researchers and by experienced leaders of large experimental units (especially in the pharmaceutical industry) who want to extract value from complete genome sequencing and other high-throughput activities for research in their field of life sciences. Bioinformatics is considered a service effort to store and retrieve biological information, to create integrated software solutions and to apply existing suites of programs in a routine manner and to supply the necessary output immediately after a request from an experimentalist is issued. This view received additional backing with the availability of WWW-servers for sequence analyses (especially BLAST servers) that appear useable like TV-sets, without necessarily understanding of the algorithms applied.

Organizationally, such an attitude results in a part-time bioinformatics person equipped with a PC linked to the internet in academia or a small service group with a high budget for ever larger and complete database and software license purchases in industry. Often, this personnel becomes dissatisfied after being overloaded with a large number of scientifically and methodically disconnected requests when, at the same time, they have to maintain their local working environment themselves. Of course, it is a serious limitation that not all necessary methods are available at WWW-servers; many respond slowly or not at all and most have to be navigated individually which causes an extensive loss of time. Not surprisingly, most nontrivial discoveries based on sequence analysis considerations have been made at other places. Specialized researchers who work at academic bioinformatics research centers with sufficient computing power and a local, well-maintained database and software environment, who have spent significant periods of their life analyzing sequence data and accumulated experience, who have taken part in methodological developments and interact with their colleagues in the field clearly have greater chances of success.[25]

Already the unfortunate notion "bioinformatics" is derived from the superficial view of doing something in biology with computers whereas the essence consists in research aimed at understanding the complex genotype-to-phenotype transformation relying on sequences, expression profiles, other types of high-throughput experimental data and the electronically available scientific literature. Only the complexity of theoretical concepts and the large amount of data require the use of computers as a tool. It is ever more important to understand that sequence analysis and computational biology comprise a field of ongoing research with many problems still unsolved. Except for obvious cases of sequence homology and sequence domain matches, considerable creativity is required for nontrivial application of existing methods to squeeze a hint on structure/function of a target considered out of its sequence. Since many functionally important sequence signatures still remain hidden, the development of new techniques recognizing more distantly related sequence family members or other biological features such as posttranslational modifications or localization signals continue to have important academic and practical relevance.

Finally, who is servicing whom? Is it more important for new biological insight to generate a new hypothesis based on theoretical data analysis or to create an experimental setup for its verification? Running simple BLAST searches or standard electrophoresis gels are equally routine (service?) activities. I think that such a discussion is counterproductive and distracts from the main issue, the scientific search coordinated among researchers with complementary professional background that strive for a common success.

The development of high-throughput experimental technologies and its first major breakthrough, the complete sequencing of the genomes of organisms ranging from viruses over bacteria, lower eukaryotes to human, has changed life science research qualitatively. For the first time in history, the biological object can be studied in its totality at the molecular level. The immediate task for the coming decade consists in assigning functions to all genes known by sequence. Apparently, sensible, quantitative gene network studies will be possible only after most of the genes have been assigned a function qualitatively and all major players of processes and their interaction topology are known. Since the new data are so large and their biological interpretation requires complex approaches, theoretical science can and must contribute decisively to research progress. Hints from theoretical studies may shorten experimental searches and save postdoc years.

Obviously, there is a new division of labour in the cognition process in life sciences. Both sides have to learn to translate prediction results into new experimental designs and to formulate new theoretical tasks based on the existing experimental data. There are now many examples of computational biology helping experimental science and more decisive support can be expected in the future. However, the interaction cannot remain a one-way street. The value of experimental work will be increasingly measured via its effect on the improvement of methods in computational biology and the increased power of extrapolation into the unknown part of the genome. For example, this includes the experimental generation of learning sets for sequence-function correlations with reliable methods to boost prediction method development, an aspect that has not received sufficient attention yet. Only an iterative interaction promises to produce maximal research progress. The experimental community should not worry: The gaps in theory are still and will remain large.

This book *Discovering Biomolecular Mechanisms with Computational Biology* unites a collection of articles by researchers in theoretical biology showing areas of life science where theory has contributed with considerable impact. Martijn A. Huynen et al, Karin Schleinkofer, Thomas Dandekar and I review methodical approaches for analyzing biomolecular sequences and structures in Section I, accompanying the text with examples of research breakthroughs. Since the scientific literature is dramatically expanding and, at the same time, becomes increasingly electronically accessible, dedicated text analysis tools for biomedical reports can link them to genomic features and contribute decisively for prioritizing research targets. Hong Pan et al and Carolina Perez-Iratxeta et al summarize the state of the art in Section II. Gene and protein network analysis is still in its infancy since many network nodes remain unknown and the quantitative characterization of most gene/protein functions is missing.

Nevertheless, studies of regulatory and metabolic networks already produce important insights with theoretical work alone, as Harmen Bussemaker and Stefan Schuster et al show in Section III. Finally, Edward N. Trifonov, Christian Schlötterer, Saurabh Astana and Shamil Sunyaev and Yuri I. Wolf et al present compelling evidence in Section IV that the evolutionary viewpoint is indispensable for understanding function and interaction of today's genes and proteins. This book will fulfill its task if the examples described here encourage the readers to find new areas in life science research where theoretical research can qualitatively change the rate of progress in understanding biomolecular mechanisms and help to move towards a quantitative and predictive biology.

References

1. Hagen JB. The origins of bioinformatics. Nat Rev Genet 2000; 1:231-236.
2. Ouzounis C. Bioinformatics and the theoretical foundations of molecular biology. Bioinformatics 2002; 18:377-378.
3. Altschul S, Boguski M, Gish W et al. Issues in searching molecular sequence databases. Nature Genetics 1994; 6:119-129.
4. Dlakic M. Chromatin silencing protein and pachytene checkpoint regulator dot1p has a methyltransferase fold. Trends Biochem Sci 2001; 26:405-407.
5. van Leeuwen F, Gafken PR, Gottschling DE. Dot1p modulates silencing in yeast by methylation of the nucleosome core. Cell 2002; 109:745-756.
6. Toth A, Ciosk R, Uhlmann F et al. Yeast cohesin complex requires a conserved protein, Eco1p(Ctf7), to establish cohesion between sister chromatids during DNA replication. Genes Dev 1999; 13:320-333.
7. Ivanov D, Schleiffer A, Eisenhaber F et al. Eco1 is a novel acetyltransferase that can acetylate proteins involved in cohesion. Curr Biol 2002; 12:323-328.
8. Bork P, Dandekar T, Diaz-Lazcoz Y et al. Predicting function: From genes to genomes and back. J Mol Biol 1998; 283:707-725.
9. Ponting CP, Schultz J, Copley RR et al. Evolution of domain families. Adv Protein Chem 2000; 54:185-244.
10. Anfinsen CB. Principles that govern the folding of protein chains. Science 1973; 181:223-230.
11. Sander C, Schneider R. Database of Homology-derived protein structures and the structural meaning of sequence alignment. Proteins 1991; 9:56-68.
12. Bork P, Gibson TJ. Applying motif and profile searches. Meth Enzymol 1996; 266:162-184.
13. Altschul SF, Koonin EV. Iterated profile searches with PSI-BLAST—a tool for discovery in protein databases. Trends Biochem Sci 1998; 23:444-447.
14. Devos D, Valencia A. Practical limits of function prediction. Proteins 2000; 41:98-107.
15. Nielsen H, Engelbrecht J, Brunak S et al. Identification of procaryotic and eucaryotic signal peptides and prediction of their cleavage sites. Protein Eng 1997; 10:1-6.
16. Eisenhaber B, Bork P, Eisenhaber F. Sequence properties of GPI-anchored proteins near the Ω-Site: Constraints for the polypeptide binding site of the putative transamidase. Protein Eng 1998; 11:1155-1161.
17. Eisenhaber B, Bork P, Eisenhaber F. Prediction of potential GPI-modification sites in proprotein sequences. J Mol Biol 1999; 292:741-758.
18. Maurer-Stroh S, Eisenhaber B, Eisenhaber F. N-terminal N-myristoylation of proteins: Prediction of substrate proteins from amino acid sequence. J Mol Biol 2002; 317:541-557.
19. Eisenhaber F, Eisenhaber B, Maurer-Stroh S. Prediction of Post-translational modifications from amino acid sequence: Problems, pitfalls, methodological hints. In: Andrade MM, ed. Bioinformatics and Genomes: Current Perspectives. Wymondham: Horizon Scientific Press, 2003:81-105.
20. Jones S, Sgouros J. The cohesin complex: Sequence homologies, interaction networks and shared motifs. Genome Biol 2001; 2, (RESEARCH0009).
21. Neuwald AF, Hirano T. Heat repeats associated with condensins, cohesins, and other complexes involved in Chromosome-related functions. Genome Res 2000; 10:1445-1452.
22. Panizza S, Tanaka T, Hochwagen A et al. Pds5 cooperates with cohesin in maintaining sister chromatid cohesion. Curr Biol 2000; 10:1557-1564.
23. Ponting CP, Benjamin DR. A novel family of Ras-binding domains. Trends Biochem Sci 1996; 21:422-425.
24. Kalhammer G, Bahler M, Schmitz F et al. Ras-binding domains: Predicting function versus folding. FEBS Lett 1997; 414:599-602.
25. Ouzounis C. Two or three myths about bioinformatics. Bioinformatics 2000; 16:187-189.

26. Iyer LM, Aravind L, Bork P et al. Quoderat demonstrandum? The mystery of experimental validation of apparently erroneous computational analyses of protein sequences. Genome Biol 2001; 2, (RESEARCH0051).
27. Yuan YP, Schultz J, Mlodzik M et al. Secreted Fringe-like signaling molecules may be glycosyltransferases. Cell 1997; 88:9-11.
28. Tatusov RL, Altschul SF, Koonin EV. Detection of conserved segments in proteins: Iterative scanning of sequence databases with alignment blocks. Proc Nat Acad Sci USA 1994; 91:12091-12095.
29. Altschul SF, Madden TL, Schaffer AA et al. Gapped BLAST and PSI-BLAST: A new generation of protein database search programs. Nucleic Acids Res 1997; 25:3389-3402.
30. Irvine KD, Wieschaus E. Fringe, a Boundary-specific signaling molecule, mediates interactions between dorsal and ventral cells during drosophila wing development. Cell 1994; 79:595-606.
31. Wu JY, Wen L, Zhang WJ et al. The secreted product of xenopus gene lunatic fringe, a vertebrate signaling molecule. Science 1996; 273:355-358.
32. Panin VM, Papayannopoulos V, Wilson R et al. Fringe modulates Notch-ligand interactions. Nature 1997; 387:908-912.
33. Fleming RJ, Gu Y, Hukriede NA. Serrate-mediated activation of notch is specifically blocked by the product of the gene fringe in the dorsal compartment of the drosophila wing imaginal disc. Development 1997; 124:2973-2981.
34. Klein T, Arias AM. Interactions among delta, serrate and fringe modulate notch activity during drosophila wing development. Development 1998; 125:2951-2962.
35. Moloney DJ, Panin VM, Johnston SH et al. Fringe is a glycosyltransferase that modifies notch. Nature 2000; 406:369-375.
36. Bruckner K, Perez L, Clausen H et al. Glycosyltransferase activity of fringe modulates Notch-delta interactions. Nature 2000; 406:411-415.
37. Nasmyth K. Separating sister chromatids. Trends Biochem Sci 1999; 24:98-104.
38. Ciosk R, Zachariae W, Michaelis C et al. An ESP1/PDS1 complex regulates loss of sister chromatid cohesion at the metaphase to anaphase transition in yeast. Cell 1998; 93:1067-1076.
39. Uhlmann F, Lottspeich F, Nasmyth K. Sister-chromatid separation at anaphase onset is promoted by cleavage of the cohesin subunit Scc1. Nature 1999; 400:37-42.
40. Chen JM, Rawlings ND, Stevens RA et al. Identification of the active site of legumain links it to caspases, clostripain and gingipains in a new clan of cysteine endopeptidases. FEBS Lett 1998; 441:361-365.
41. Eichinger A, Beisel HG, Jacob U et al. Crystal structure of gingipain R: An Arg-specific bacterial cysteine proteinase with a Caspase-like fold. EMBO J 1999; 18:5453-5462.
42. Uhlmann F, Wernic D, Poupart MA et al. Cleavage of cohesin by the CD clan protease separin triggers anaphase in yeast. Cell 2000; 103:375-386.
43. Jenuwein T, Laible G, Dorn R et al. SET domain proteins modulate chromatin domains in Eu- and heterochromatin. Cell Mol Life Sci 1998; 54:80-93.
44. Aagaard L, Laible G, Selenko P et al. Functional mammalian homologues of the drosophila PEV-modifier Su(Var)3- 9 encode Centromereassociated proteins which complex with the heterochromatin component M31. EMBO J 1999; 18:1923-1938.
45. O'Carroll D, Scherthan H, Peters AH et al. Isolation and characterization of Suv39h2, a second histone H3 methyltransferase gene that displays Testis-specific expression. Mol Cell Biol 2000; 20:9423-9433.
46. Klein RR, Houtz RL. Cloning and developmental expression of pea ribulose-1,5-bisphosphate Carboxylase/Oxygenase large subunit N-methyltransferase. Plant Mol Biol 1995; 27:249-261.
47. Rea S, Eisenhaber F, O'Carroll D et al. Regulation of chromatin structure by Site-specific histone H3 methyltransferases. Nature 2000; 406:593-599.
48. Lachner M, O'Carroll D, Rea S et al. Methylation of histone H3 lysine 9 creates a binding site for HP1 proteins. Nature 2001; 410:116-120.
49. Bannister AJ, Zegerman P, Partridge JF et al. Selective recognition of methylated lysine 9 on histone H3 by the HP1 chromo domain. Nature 2001; 410:120-124.

SECTION I

Deriving Biological Function of Genome Information with Biomolecular Sequence and Structure Analysis

CHAPTER 1

Reliable and Specific Protein Function Prediction by Combining Homology with Genomic(s) Context

Martijn A. Huynen,* Berend Snel and Toni Gabaldón

Abstract

Completely sequenced genomes and other types of genomics data provide us with new information to predict protein function. While classical, homology-based function prediction provides information about a proteins' molecular function (what does the protein do at a molecular scale?), the analysis of the sequence in the context of its genome or in other types of genomics data provides information about its functional context (what are the proteins' interaction partners, and in which biological process does it play a role?) Genomic context data are however inherently noisy. Only by combining different types of genomic(s) context data (vertical comparative genomics) or by combining the same type of genomics data from different species (horizontal comparative genomics) do they become sufficiently reliable to be used for protein function prediction. Homology-based function prediction and context-based function prediction provide complementary information about a protein's function and can be combined to make predictions that are specific enough for experimental testing. Here we discuss the genomic coverage and reliability of combining genomics data for protein function prediction and survey predictions that have actually led to experimental confirmation. Using a number of examples we illustrate how combining the information from various types of genomics data can lead to specific protein function predictions. These include the prediction that the Ribonuclease L inhibitor (RLI) is involved in the maturation of ribosomal RNA.

Introduction

Genome sequencing provides us with an abundance of genes whose functions are not determined experimentally and have to be predicted by bioinformatics. The classic tool to do so, homology detection, is mainly suited to predict the molecular function of a protein. Having complete genome sequences we would also like to know protein function at a higher level, like the pathway or complex a protein belongs to.[1] Bioinformatics supplies us with a growing number of so-called genomic context methods that exploit the genomics data themselves to predict such interactions. These methods exploit the fact that the genes of functionally interacting proteins tend to be associated with each other in genomes or in other types of genomics data. At the level of genome sequences, gene fusion,[2,3] the conservation of gene order,[4,5] the co-occurrence of genes among sequenced genomes,[6,7] or genes having a

*Corresponding Author: Martijn A. Huynen—Nijmegen Center for Molecular Life Sciences, p/a Center for Molecular and Biomolecular Informatics, Toernooiveld 1, 6525 ED Nijmegen, Netherlands. Email: huynen@cmbi.ru.nl

Discovering Biomolecular Mechanisms with Computational Biology, edited by Frank Eisenhaber.

complementary distribution,[8] the sharing of regulatory elements,[9-11] and methods that use sequence information of the proteins itself.[12-14] have been proposed and implemented (Fig. 1). These methods have in common that they exploit the availability of multiple sequenced genomes to increase an in itself weak signal for functional interaction between proteins. To give one example, that two genes are neighbors in one bacterial species is only a weak signal that they functionally interact, but when that gene order is conserved among many genomes it does become a strong signal, even of physical interaction between the proteins.[15] Genomic-context methods are becoming well established and have been the subject of many reviews already.[8,16-18] Presently the focus is on combining and integrating them with each other and with other types of genomics data. As is the case of methods based on comparing genomes, also using the evolutionary conservation of coexpression,[19-21] or of physical interaction as measured with yeast-2-hybrid[19,22,23] leads to a drastic increase in the reliability of the results. Another way of combining genomics data, detecting the same interaction in different types of data, e.g., two genes tend to be coexpressed in *Saccharomyces cerevisiae* and their proteins interact in a yeast-2-hybrid experiment in the same species, increases the likelihood of that interaction.[24,25] In this book chapter we will focus on the practical applicability of these methods: can we use these tools to make predictions that are not only reliable, but that are also specific enough to design experiments to test them, and therewith complete a research "circle" from (genomics) experiment to theory to experiment?

We survey the predictions that were actually experimentally verified and make a number of new ones. The latter will illustrate that, notwithstanding all the advances in making the data and tools available on the web, making specific predictions still requires manual intervention and creativity to integrate the different types of information and make specific predictions.

Types of Genomic Context

Gene Fusion

The finding of two or more proteins encoded by separate genes of which orthologs in a different species are encoded in a single gene (Fig. 1A), reveals a gene fusion or gene fission event.[26] This is the most direct form of genomic context and, from a functional point of view, the fusion of two proteins can result in an enhancement of the interaction between their respective biochemical activities to facilitate, for example, the channelling of a substrate.[27] Using this approach to predict functional interactions in complete genomes was introduced in 1999 by Marcotte et al[2] and Enright et al.[3] In concordance with the above mentioned substrate channelling effect, most of the observed fusions events involve metabolic enzymes, although the fusions do not always involve subsequent steps in the pathway.[3,28] In *Escherichia coli* three quarters of the total of gene fusions affect metabolic genes.[29]

Conservation of Gene Order in Prokaryotes

The first pairwise genome-wide sequence comparisons revealed that even closely related species lack large scale conservation of gene order,[7,30-32] indicating that in the course of evolution genomes are rapidly rearranged and shuffled. Yet in prokaryotes some clusters of genes appear conserved in evolution (Fig. 1B), including the relative location of the genes within them, over large evolutionary distances. Further inspection of these genes revealed that they tend to encode proteins that functionally interact,[5,33] and that they tend to be part of the same operon.[34] As in the case of gene fusion, since conservation of chromosomal proximity has functional meaning it can be used to predict functional interaction between the components of conserved gene clusters. This was proposed in 1998 by Overbeek et al[33,35] and Dandekar et al,[5] by measuring conservation of genes in runs (sets of genes encoded in the same strand and separated by less than 300 bases) and conservation of neighboring genes respectively. Although there are some hints of chromosomal clustering of functionally interacting genes in eukaryotes, e.g., in polycistronic transcripts in Nematodes[36] these do not appear strong enough to predict functional interactions with any level of confidence.

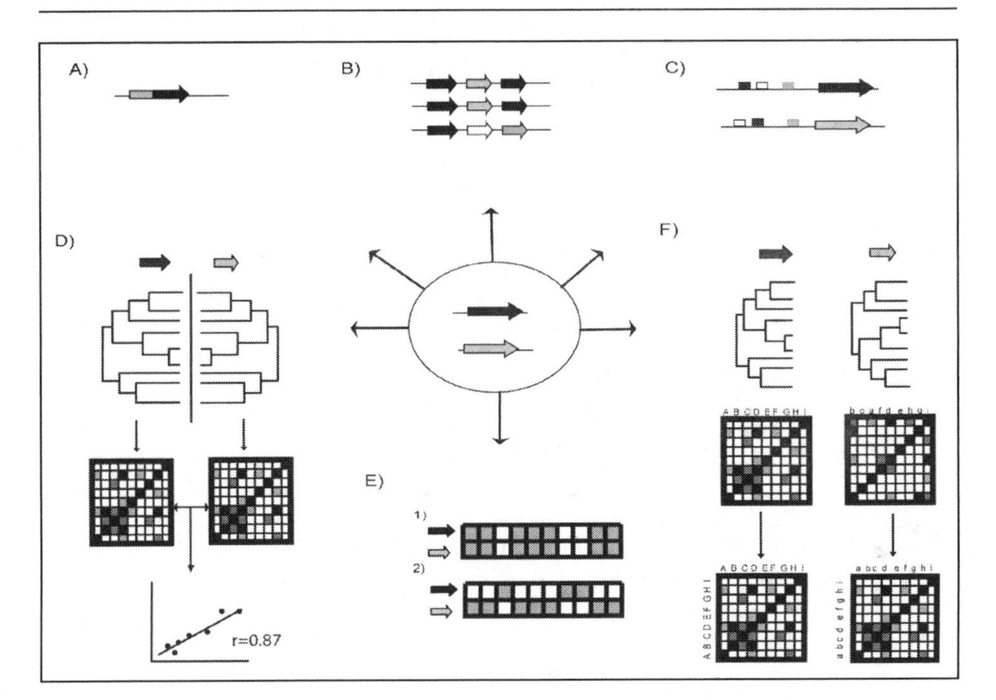

Figure 1. Types of genomic context between two genes of a certain genome (centre of the figure) that indicate a functional interaction between the encoded protein. A) Gene fusion: the proteins are encoded in a single gene in another genome. B) Gene-order conservation: the genes are close in the chromosome and transcribed in the same direction in several, distantly related species. C) Sharing of a regulatory sequence: the genes share a regulatory sequence indicating coregulation. D) 1. Similar evolutionary pattern: the genes have a similar pattern in terms of the distance matrix of the homologs. E) Phylogenetic pattern: 1) co-occurrence: the genes have a similar pattern of presence/absence across species or 2) complementary distribution: the genes have an anti-correlated pattern in their presence across species. F) Matrix alignment (similar to D): to predict, from two protein families, which protein from family A interacts with which protein from family B one can align the distance matrices of the two families with each other. After alignment, proteins in corresponding columns are predicted to interact.

Sharing of a Regulatory Sequence

In eukaryotes, but also in prokaryotes, genes do not have to be neighbors to be coregulated: they can be part of the same regulon without being part of the same operon by sharing the same transcription factor binding sites (Fig. 1C). Prediction of coregulation based on comparative genome analysis is being used to predict protein function.[37] Comparative genome analysis does play an even more essential role here than in gene order conservation or gene fusion, as it can help in determining what the regulatory sequences are in the first place, by identifying conserved DNA sequences upstream of orthologous genes. These regulatory sequences can then subsequently be searched for in complete genomes to identify regulons,[38] and therewith potentially interacting proteins.

Co-Occurrence of Genes in Genomes

Although the fact that two genes are encoded together in one genome provides only a very weak signal that they to interact, when they are encoded in a considerable number of genomes, and are both absent from others this signal becomes strong enough for function prediction (Fig. 1E). This technique, called gene co-occurrence or phylogenetic profiles, was proposed[6,7] and verified by studies showing that proteins with a similar distribution across species have a high tendency to functionally interact.[6,7,39,40] The use of gene co-occurrence to predict protein interactions is continuously undergoing technical improvements, among others to filter out the phylogenetic bias in the sequenced genomes by using evolutionary information to measure the distance between profiles[41] or collapsing into a single node parts of the profile that represent related species that share the presence or absence of a certain gene.[42] A reverse use of phylogenetic profiles to predict function is the identification of proteins with complementary or anti-correlated profiles[8] (Fig. 1E) to detect nonorthologous gene displacements.[43] In general, the detection of nonorthologous gene displacement by complementary phylogenetic profiles is combined with gene-order conservation to increase the signal: i.e., does the "new" gene occur in conserved operons with the other genes with which it is supposed to interact, replacing the old gene not only in terms of functional context but also in terms of genomic context.

Coevolution of Sequences

Another variant of the use of coevolution to predict protein interaction uses the evolutionary information that is contained at the level of the sequences themselves. For specific cases of protein families known to interact, such as insulin and its receptors[44] or the chemokine-receptor system,[14,45] their phylogenetic trees are relatively similar compared to other protein families (Fig. 1D). Valencia and colleagues[12,46] made use of this property to search for interaction partners within the *E. coli* proteome by measuring the correlation between the distance matrices used to build the phylogenetic trees. Ramani and Marcotte[13] used a similar approach to predict the binding specificities among members of 18 ligand and receptor families that posses many paralogs in the human genome (Fig. 1F). The coevolution of interacting partners can be followed more closely by searching for mutations that are correlated in both protein families (they occur in the same species), these positions may correspond to residues on the interface that undergo compensatory mutations in one protein to compensate the effects of mutations in the other. This method has been used for the prediction of interacting partners based on the finding of pairs of proteins with correlated mutations.[47,48] It has the advantage that provides not only the prediction of interacting partners but also of potentially interacting residues.

Accuracy and Genomic Coverage of Context Based Predictions

Analyses of the reliability of genomic context methods to predict functional interactions indicate that it is generally high, specifically for gene fusion (72%)[28,49,50] and for gene-order conservation (80%).[49-51] One should keep in mind however that the benchmarks that are used to estimate this reliability are often quite general: proteins are regarded as interacting when they have a similar set of SWISS-PROT keywords,[52] or fall on the same metabolic map in KEGG,[53] and thus the predictions that can be made tend to be general too. The availability of genomics-scale yeast-2-hybrid[54] or identification of protein complexes by mass spectrometry analyses[55] should allow more systematic benchmarking of the genomic context methods for the prediction of physical interaction, were it not that these data themselves are not always of high quality: By comparing experimental genomics' techniques, mRNA correlated expression, and genomic context predictions to a classic set of "trusted" physical interactions, it was shown that genomic context predictions actually had both a higher coverage and a higher accuracy than not only mRNA coexpression, but also than direct experimental techniques like yeast-2-hybrid or high-throughput mass-spectrometric protein complex identification (HMS-PCI).[24] As the combination of genomic context data with experimental data increases the reliability of the predictions, genomic context can also be used as a filter, to improve the quality of the experimental data,[24,56] albeit at a loss of coverage.

Genomic-context based predictions cover a large fraction of the genome. Based on gene-order conservation, gene fusion and gene co-occurrence we can presently predict with 80% confidence functional links for the majority of the proteome of prokaryotes (64% in *Mycoplasma genitalium* and 60% in *E.coli*) and for a substantial fraction of the proteome of the eukaryote *Saccharomyces cerevisiae* (26%).[57] It should thereby be noted that some hypothetical proteins with a significant genomic context are only linked to other hypothetical proteins. Links between hypothetical proteins cannot be used for function prediction, but they are relevant because they provide information about the topology of the network of interactions in a cell. Using genomic context we can thus already obtain a view on the network of interactions within a cell, even if we do not know the functions of all the individual elements of that network.

Experimentally Verified Context Predictions

Applicability of genomic context methods for protein function prediction can, in the long run, only be established by experimental confirmation of their predictions. We have identified 22 cases where functional interactions and function were predicted to a varying level of specificity and either published before or along with the experimental verification (Table 1). In these cases gene fusion, gene-order conservation, gene co-occurrence, a complementary distribution of genes over species, and the sharing of regulatory elements have been used successfully to predict new protein functions. Gene-order conservation contributes the largest share of the predictions (Fig. 2), covering on its own more than half of all successful predictions. Its use is however limited to genes that (also) occur in prokaryotes. Note that using genomic context methods to design an experiment is not trivial because the indications of function or of functional interaction are not very specific. The methods do not predict what the type of interaction between the proteins is, it could e.g., be regulatory, physical or being part of the same pathway or process (Table 1).[49] The exception to this is

Table 1. Experimental verification of context predictions

Protein/Gene	Context	Type of Interaction	Function
Mt-K	gene order	physical interaction	double-stranded DNA repair[87]
GnlK	gene order	physical interaction	signal transduction for ammonium transport[88,89]
PH0272	gene order	metabolic pathway	methylmalonyl-CoA racemase[90]
PrpD	gene order	metabolic pathway	2-methylcitrate dehydratase[49,91]
arok	gene order	metabolic pathway	shikimate kinase[92]
ComB	gene order	metabolic pathway	2-phosphosu-lfolactate phosphatase[93]
KynB	gene order	metabolic pathway	kynurenine formamidase[94]
PvlArgDC	gene order	metabolic pathway	arginine decarboxylase[95]
FabK	gene order	metabolic pathway	enoyl-ACP reductase[96]
FabM	gene order	metabolic pathway	trans-2-decenoyl ACP isomerase[97]
COG0042	gene order	tRNA modification	tRNA-dihydrouridine synthase[98]
Yfh1	co-occurrence	process	iron-sulfur protein maturation[99,100]
YchB	co-occurrence	metabolic pathway	terpenoid synthesis[101]
SmpB	co-occurrence	process	trans-translation[6,102]
ThyX	complement	enzymatic activity	thymidilate synthase[8,103]
ThiN	complement	enzymatic activity	thiamine phosphate synthase[37,58]
Prx	fusion	pathway	peroxiredoxin[104]
YgbB	fusion/order	metabolic pathway	terpenoid synthesis[105]
SelR	fus./ord./co-o	enzymatic activity	methionine sulfoxide reductase[8,49,106]
FadE	reg. sequence	metabolic pathway	acyl CoA dehydrogenase[107,108]
TogMNAB	reg. sequence	metabolic pathway	Oligogalacturonide transport[109,110]
MetD	reg. sequence	metabolic pathway	Methionine transport[79,111]

In all cases genomic context was used to predict a functional interaction between proteins, and was this interaction subsequently experimentally verified. In the cases where more than one reference is given the functional link was published separately and before the experimental verification.

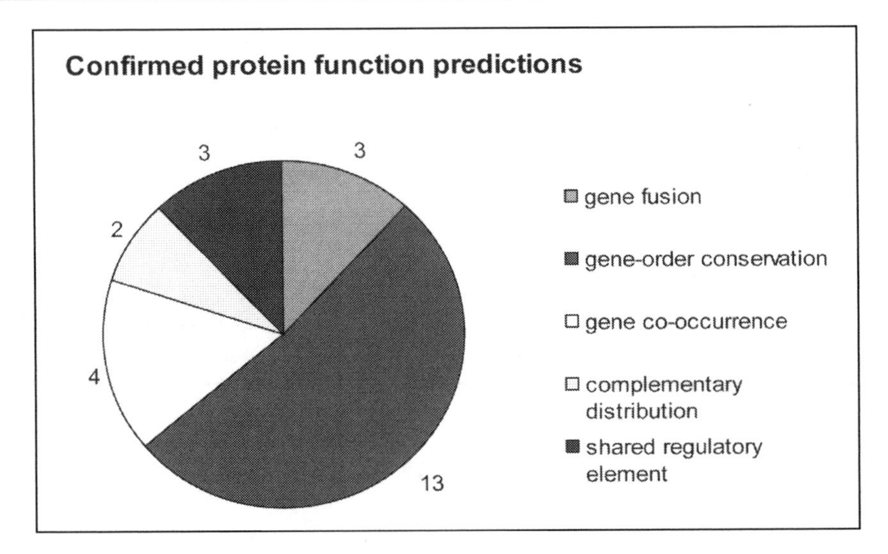

Figure 2. Quantitative coverage of various types of genomic context in terms experimentally confirmed protein functions that were predicted by various genomic context methods. Gene-order conservation alone covers 50% of the confirmed predictions, the remaining ones are more or less equally divided among the other types of genomic context. The data are from Table 1.

when genes have a complementary phylogenetic distribution. In that case the proteins, rather than just interact, should actually have the same function as they replace each other in the genome and the function prediction can be specific, provided that the function of one of the proteins is known.[58]

One way to increase the prediction specificity is to include the degree to which the genomic context is conserved. The stronger the evolutionary conservation of a genomic context pattern (e.g., the more often that the genes are neighbors), the more likely that the proteins not only functionally interact, but also that they interact in the most direct way: i.e., by being involved in the same reaction and forming a protein complex.[15] Another promising direction of research to increase the specificity of the predictions is to include the local topology of the network. Locally densely connected networks reflect physical complexes, while less connected ones correlate with signaling pathways.[55] Generally however, it is left to the researcher to combine the genomic context information with data on homology relations and with data on e.g., the phenotypic effects of deletion of the protein or on missing steps in a pathway, to make a specific, testable prediction about the protein's function, see also reference 38. We will illustrate these principles with a number of examples of using and combining various types of genomic information to arrive at specific predictions.

Practical Examples of New Protein Function Predictions Based on Genomics Data

A Hypothetical Protein Involved in Bacterial DNA Repair

The orthologous group of proteins that has been classified as COG0718 in the COG database (www.ncbi.nlm.nih/COG),[59] and is present in virtually all sequenced Bacterial genomes has at the sequence level no detectable homolog with known function. A structure has recently been determined,[60] but also by comparisons at the structure level no homology could be detected. In such a situation, although a hypothesis about a specific function is hard to obtain, genomic context can at least pinpoint the biological process in which the

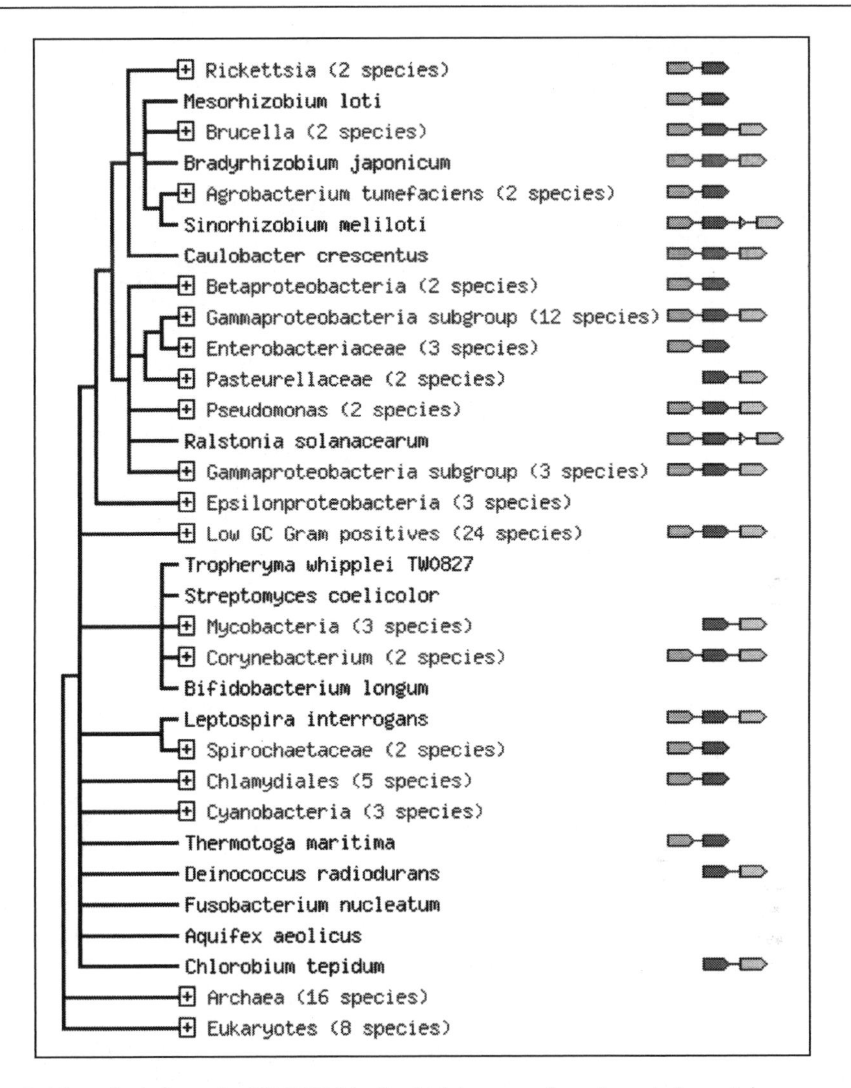

Figure 3. A hypothetical protein, COG0718 (red), which is present throughout the bacteria has a conserved gene order with the recombinational repair protein RecR (orange) and with DNA polymerase subunits gamma and tau, encoded by de DnaX gene (brown). Such strong conservation of gene order tends to indicate a physical interaction between the encoded proteins, implicating COG0718 in recombinational repair. Data from STRING (http://string.embl.de). A color version of this figure is available online at www.Eurekah.com.

protein plays a role. Examination of the STRING database (http://string.embl.de) shows that the genes from COG0718 have a strong conservation of gene order with both the recombinational repair protein RecR and the DNA polymerase subunits gamma and tau (encoded by a single gene, dnaX) (Fig. 3). This is consistent with the finding that they are cotranscribed,[61] and suggests a role for COG0718 in recombinational repair.[49] Although having the structure of the protein does in this case not give information via homology to proteins with known functions, Lim and coworkers[60] did note that the conserved, negatively charged residues of this orthologous protein do cluster on one side of the protein. They suggest that the protein could have a regulatory role, with the negatively charged residues competing with the DNA for the binding of DNA-binding proteins in the DNA replication fork repair process.[60]

FMN Binding Proteins in Trehalose Metabolism

An example of a case in which we can use both sequence homology as well as conservation of coexpression for function prediction is the *S.cerevisiae* hypothetical gene YBR052c. Examination of the SMART domain detection tool (smart.embl.de)[62] indicates that YBR052C is a hypothetical protein that contains a flavin mononucleotide (FMN) binding domain known to be involved in redox-reactions. YBR052c is coexpressed with YDR074w/TPS2 (trehalose-6-phosphate synthase) in a large set of yeast experiments (r, the uncentered correlation coefficient of their expression levels in expression data of Hughes and coworkers[63] is larger than 0.6). Furthermore, this coexpression is conserved: both proteins have a homolog in *S.cerevisiae*, and these homologs, YCR004C and TPS1, are again coexpressed with each other ($r > 0.6$). This type of evolutionary conservation coexpression after parallel gene duplication leads to a similar increase in the reliability as the conservation of coexpression after speciation.[20] YBR052C and YCR004C are thus likely to play a role in trehalose metabolism, possibly, given their FMN binding capacity, in its role as antioxidant.[64]

Interaction between Bola and a Mono-Thiol Glutaredoxin in Oxidative Stress

The orthologous group COG0271 that contains the *Escherichia coli* protein Bola[65] has a wide phylogenetic distribution, including almost all alpha, beta and gamma-proteobacteria and eukaryotes, including *Homo sapiens* (Fig. 4A). Although the molecular function of Bola is not known, it has been implicated in cell division,[65] and is expressed under stress conditions.[66] A role in defense against (oxidative) stress is supported by data on an *Schizosaccharomyces pombe* ortholog of BolA, UVR31, that is upregulated under UV radiation.[67] The conservation of gene order with a monothiol glutaredoxin-like protein (COG0278), as well as an almost identical phylogenetic distribution with the monothiol glutaredoxin-like protein indicate an interaction between the two (Fig. 4A,B). Interaction with a monothiol glutaredoxin-like protein is also observed in genomics protein interaction data: examination of the BIND database (www.blueprint.org/bin/bind.php) indicates that in *S.cerevisiae* the ortholog of Bola and an ortholog of the glutaredoxin-like protein GRX3 have been observed to interact in a yeast-2-hybrid screen,[68] as well as in Tandem Affinity Purification.[55] Furthermore, this interaction is conserved, it is also present in the yeast-2-hybrid screens of *Drosophila melanogaster.*[69]

These data all indicate that Bola interacts with a mono-thiol glutaredoxin in its role in defense against oxidative stress. The role of this interaction cannot be revealed from genomic-context data, a hint does however come from homology. The structure of Bola has recently been described,[70] and it is sufficiently similar to that of another protein involved in defense against oxidative stress, Osmc,[70] that they are classified in the same fold in the DALI protein structure database.[71] OsmC is a reductase that has been proposed to use cysteine thiol groups to reduce substrates that cause or result from oxidative stress.[70] Bola does not have any conserved thiol groups. Combining context information about the interaction with the mono-thiol glutaredoxin with the homologous relation to OsmC suggests that Bola acts as a reductase, using the thiol group of mono-thiol glutaredoxin as reducing equivalents.

COG2835 Is Involved in Lipopolysaccharide Synthesis

For the predicted human protein MGC14156 (Refseq: NP_116295) no functional characterized homologous protein can be detected by homology searches. Besides two sequences in mouse, the most similar homologs of this protein are bacterial sequences that belong to the cluster of orthologous genes COG2835. A hint about the proteins' location in the cell comes from the protein's N-terminus where a mitochondrial-targeting signal is detected. The mitochondrial localization of this protein is supported with high probabilities by different targeting predictors such as MITOP (0.99), Target-P (0.96) or Predotar (0.99). Moreover, the

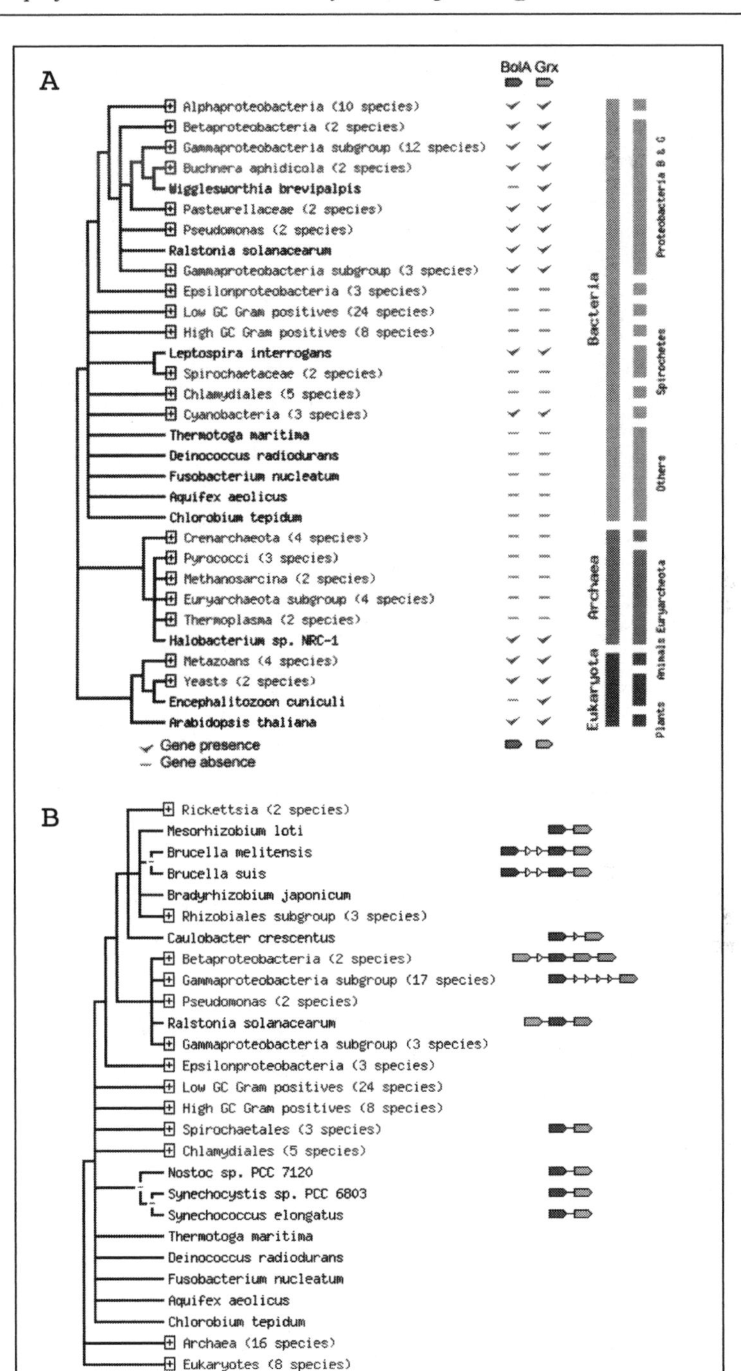

Figure 4. Bola (red), a stress induced-morphogen whose molecular function is unknown has a strong genomic association with a mono-thiol glutaredoxin (green). It has both an almost identical phylogenetic distribution with the mono-thiol glutaredoxin (A) as a well as a strong gene-order conservation (B). This indicates a functional interaction between the proteins. A color version of this figure is available online at www.Eurekah.com.

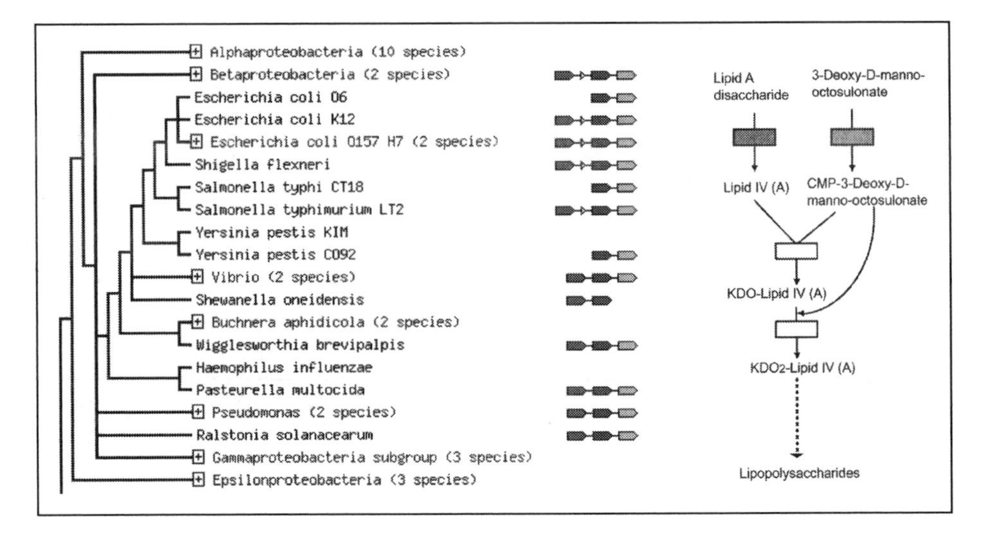

Figure 5. COG2835, a hypothetical protein whose molecular function is unknown has a conserved gene order with two genes involved in lipo-polysaccharide synthesis, COG1663 (Tetraacyldisaccharide-1-P 4-Kinase) and COG1212 (CMP-2-keto-3-deoxyoctulosonic acid synthetase. This implicates a role of COG2835 in lipo polysaccharide synthesis as well. The lipopolysaccharide synthesis pathway is on the right.

alpha-proteobacterial origin of this protein, determined by phylogenetic analyses,[72] suggests that its mitochondrial localization has an ancient origin. The genome context analysis of COG2835 carried out on the STRING server (Fig. 5) reveals a significant association with two other orthologous groups: COG1663 (Tetraacyldisaccharide-1-P 4-Kinase) and COG1212 (CMP-2-keto-3-deoxyoctulosonic acid synthetase). These two orthologous groups have enzymatic activities acting in the same biochemical pathway: the synthesis of lipopolysaccharides (LPS) (Fig. 5). The strong association observed in terms of conserved gene order and neighborhood with two genes involved in the same pathway strongly supports an implication of COG2835 in lipopolysaccharyde synthesis or metabolism in bacteria. Its role in *Homo sapiens* is less clear, as that species does not synthesize lipopolysaccharides.

RNase L Inhibitor in Ribosome Biogenesis

In the final example we will show a wealth of evidence from genomic context data and homology relations suggesting that the protein RNase L inhibitor (RLI) is involved in ribosome biogenesis through an interaction with ribosomal RNA. In *H. sapiens* RLI has been implicated in the 2'-5' oligoadenylate pathway, an interferon inducible RNA degradation pathway responsible for many of the antiviral and antiproliferative effects of interferon.[73] In this pathway the RLI reversibly associates with the endoribonuclease RNase L which it inhibits,[74] thus preventing the degradation of viral RNA. RLI has also been implicated in a very different activity: it is recruited by the HIV-1 protein GAG and required for capsid assembly.[75] These two reported activities of RLI, interaction with RNAse L or with HIV-1 proteins are not likely to be the whole story about RLI's function. First of all because RLI itself is present in all eukaryotes and all Archaea that have been sequenced so far (Fig. 6), but an examination of the SMART database[76] indicates that only mammals have proteins with the domain organization of its interaction partner in human, RNase L. Secondly because a role in the assembly of HIV-1 capsids can hardly be the original reason of the proteins existence outside HIV genomes. We therefore examined information from genomic context data and from homology to derive a hypothesis about the function of RLI besides the roles described above.

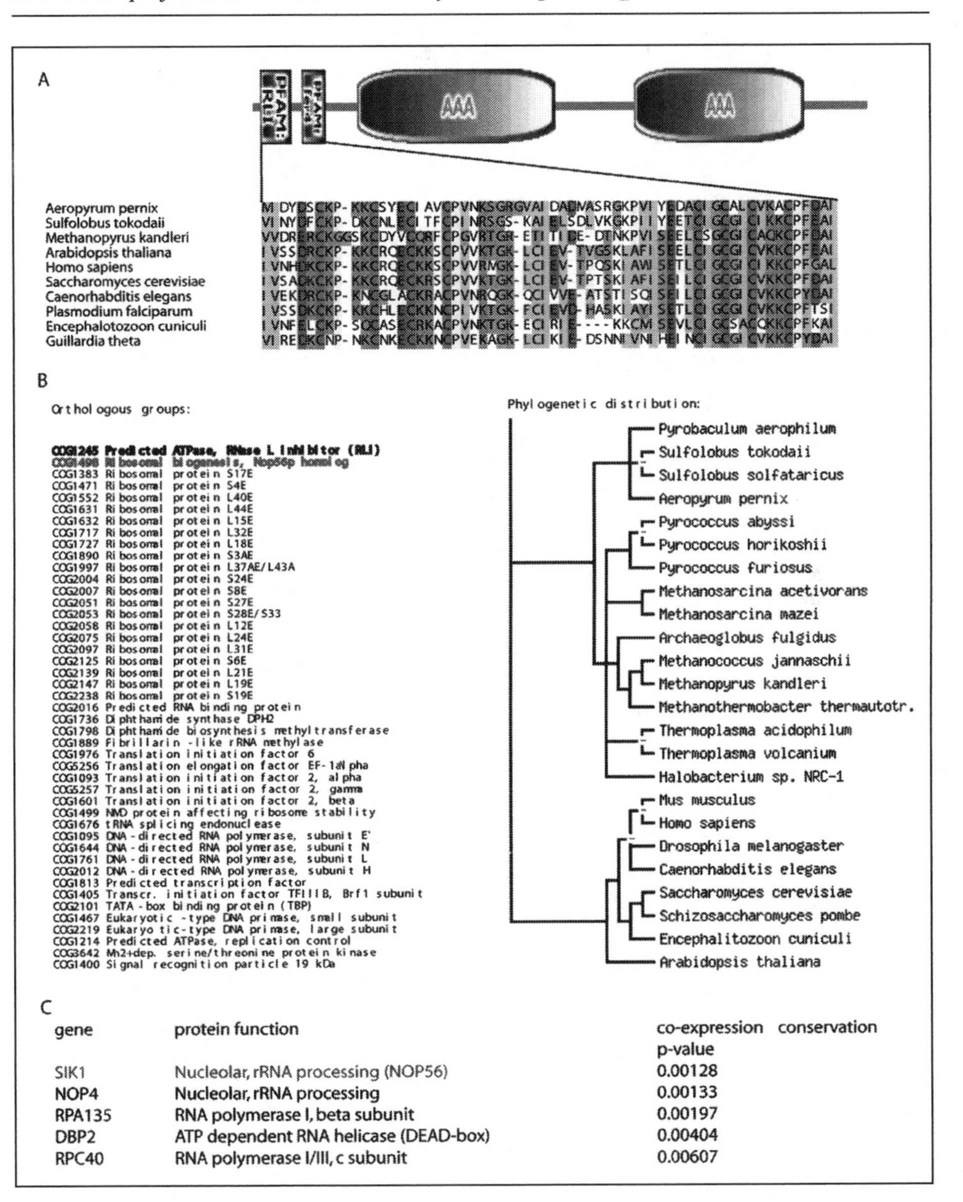

Figure 6. A combination of genomic context and sequence data implicate the RNAse L inhibitor in ribosome synthesis, specifically in ribosomal RNA maturation. A) The domain organization of RLI from SMART (smart.embl.de), and an alignment of the N-terminal, cysteine rich domains of a representative set of sequences, constructed with clustalx. In each sequence both sets of 4 cysteines contain at least one intercysteine loop with a positively charged residue (Lysine or Arginine). B) The phylogenetic distribution of RLI from STRING (string.embl.de): the orthologous group is present in all (sequenced) Archaea and Eukaryotes and is absent from the Bacteria. C) Conserved coexpression from (cmgm.stanford.edu/~kimlab/multiplespecies): The genes that are conservedly coexpressed with RLI (P value < 0.001) and for whom functional information is available can all be linked to transcription and processing of rRNA. Reprinted from reference 113, with permission.

Co-Occurrence of RLI with Ribosome (Biogenesis) Genes

First we examined whether there are other orthologous groups that specifically tend to cooccur with RLI in genomes, indicating a possible functional interaction between the proteins. An examination of the SMART tool (http://www.smart.embl.de) indicates that RLI contains two 4-cysteine domains and two ATPse domains (Fig. 6A), and indicates that genes with an identical domain composition as RLI, although widespread in evolution, only occur once per species. There appears not to have been any gene duplication in the evolution of this family, making RLI part of a very well-defined orthologous group. Indeed in the COG database as implemented in STRING (http://www.string.embl.de) indicates all proteins with the RLI domain composition are part of the same orthologous group. Further examination indicates that only 55 orthologous groups have a phylogenetic distribution that is identical to RLI (Fig. 6B). For the orthologous groups in this set of which we know the function (44), nearly all are either involved in translation or ribosome biogenesis 33 (60%), in transcription 7 (13%) and in DNA replication, recombination and repair 3(5%). These correlations point to a role of RLI in DNA duplication/transcription or RNA processing.

Conserved Coexpression of RLI with Ribosome (Biogenesis) Genes

From a second type of genomic context, the conservation of coregulation,[20,77] comes a prediction that is consistent with this observation, but that is more specific. Among four species from which extensive RNA-expression data are available (*S.cerevisiae, C.elegans, D. melanogaster, H. sapiens)* RLI is conservedly coexpressed with a number of proteins involved in the transcription or processing of rRNA (Fig. 6C), data from the conserved coexpression web-site (http:/cmgm.Stanford.edu/cgi-bin/cgiwrap/kimlab/ multiplespecies/search.pl). Some of the proteins that are conservedly coexpressed with RLI in four species have an identical phylogenetic distribution to it over all sequenced genomes. They include the nucleolar protein SIK1 (NOP56) from yeast that is involved in rRNA methylation[20]) and the B and C subunits of RNA polymerase I. The combination of gene co-occurrence and conserved coexpression indicates a role of RLI in the ribosome or ribosome biogenesis, specifically in the processing of ribosomal RNA.

Physical Interaction of RLI with HCR1

Large-scale analyses of physical interaction between proteins have been published for a couple of species, and also here one observes an increase in the likelihood that proteins interact when that interaction has been observed in multiple species.[22,23] For RLI no evolutionary conserved physical interaction can be observed however. Within *S. cerevisiae* another type of "conservation" is however present. As indicated by the GRID database (biodata.mshri.on.ca/ yeast_grid), two independent physical interaction detection techniques, yeast-2-hybrid[78] as well as Affinity Purification directed towards ribosome biogenesis[79] detect interaction of RLI with the same protein, HCR1. Consistent with the observations above, HCR1 is involved in the processing of ribosomal RNA,[80] as well as in translation initiation.[80]

Cytoplasmatic Localization of RLI

Data on the location of a protein can be indicative of interactions with other proteins in the case that there are few proteins in that compartment,[81] and the probability that they do interact is thus relatively large. In *S. cerevisiae* RLI1 appears to be located in the cytoplasm, which is, given the large number of proteins in that cell compartment not indicative of any specific pathway. For a role in ribosome biogenesis, as is indicated by other genomics data, localization in the nucleolus would appear more likely. In *S.cerevisiae*, the last steps of ribosome biogenesis do however take place in the cytoplasm,[82] where RLI's interaction partner HCR1 is also located.[80] So the localization evidence, although it does not specifically support RLI's role in ribosome biogenesis, it is not inconsistent with it either.

Phenotype Data: A Delayed Slowdown of Protein Synthesis

Knockouts of RLI in yeast are lethal, and are therefore nor very informative for its function other than that RLI is essential for the cell,[83] while over expression of the protein leads to a mild inhibition of growth. Recently, inducable depletion of RLI in *Trypanosoma brucei* was established with RNAi.[84] Although one might expect that, given its role as an RNAse inhibitor, depletion of RLI would lead to decreased levels of mRNA no such effect was observed.[84] Rather, after a delay of 48 hours, protein synthesis appeared to be reduced,[84] which is consistent with a role of RLI in ribosome biogenesis.

Domain Composition of RLI Hints at Interaction with RNA

RLI does contain domains whose separate molecular functions are quite well known, although they have not been found in combination with each other. At its N-terminus the protein contains two domains with four conserved cysteines each, one unique to RLI, and one a 4Fe-4S binding domain. It furthermore contains 2 ATPase domains (Fig. 6A). One possibility to link this domain organization to the role in ribosome biogenesis that was indicated by the co-occurrence and genomic context analyses, lies in the 4Fe-4S binding domain. Aside from playing a role in redox reactions, these domains have also been observed in DNA binding proteins endonuclease III[85] and the DNA glycosylase MutY.[86] The 4Fe-4S cluster is hypothesized to stabilize the fold, presenting a loop that extends from the protein to the backbone of the DNA.[85] Consistent with a role of both the 4-cysteine domains of RLI in binding the backbone of the double stranded rRNA (negatively charged) is that they both contain conserved Lysines (positively charged), the first domain between its first two cysteines, the second one between its third and fourth cysteine.

Discussion

Examples like the ones above show the potential and limits of using comparative genomics in protein function prediction. We can pinpoint to a role in a process, but cannot always predict what exactly that role is. In the case of a metabolic pathway, a molecular function like enzymatic activity and a context function like the pathway can often be matched to obtain a specific prediction. In the case of the RNase L inhibitor case the situation is less obvious, although a lot can be gained from combining various types of genomics and nongenomics information. The idiosyncratic nature of the case stories of function prediction discussed above is a reflection of the idiosyncratic nature of biotic systems. This is not to downplay the role of bioinformatics: combining the information either along vertical lines (between different types of experiments) or along horizontal lines (between different species) such that it becomes reliable and presenting the information in a comprehensive way is essential for optimal exploitation. Turning all that information into a specific, testable function prediction does require human creativity.

References

1. Bork P, Dandekar T, Diaz-Lazcoz Y et al. Predicting function: From genes to genomes and back. J Mol Biol 1998; 283:707-725.
2. Marcotte EM, Pellegrini M, Ng HL et al. Detecting protein function and protein-protein interactions from genome sequences. Science 1999; 285:751-753.
3. Enright AJ, Iliopoulos I, Kyrpides NC et al. Protein interaction maps for complete genomes based on gene fusion events. Nature 1999; 402:86-90.
4. Overbeek R, Fonstein M, D'Souza M et al. Use of contiguity on the chromosome to predict functional coupling. In Silico Biol 1998; 1:93-108.
5. Dandekar T, Snel B, Huynen M et al. Conservation of gene order: A fingerprint of proteins that physically interact. Trends Biochem Sci 1998; 23:324-328.
6. Pellegrini M, Marcotte EM, Thompson MJ et al. Assigning protein functions by comparative genome analysis: Protein phylogenetic profiles. Proc Natl Acad Sci USA 1999; 96:4285-4288.
7. Huynen MA, Bork P. Measuring genome evolution. Proc Natl Acad Sci USA 1998; 95:5849-5856.

8. Galperin MY, Koonin EV. Who's your neighbor? New computational approaches for functional genomics. Nat Biotechnol 2000; 18:609-613.
9. Gelfand MS, Mironov AA, Jomantas J et al. A conserved RNA structure element involved in the regulation of bacterial riboflavin synthesis genes. Trends Genet 1999; 15:439-442.
10. McGuire AM, Hughes JD, Church GM. Conservation of DNA regulatory motifs and discovery of new motifs in microbial genomes. Genome Res 2000; 10:744-757.
11. van Nimwegen E, Zavolan M, Rajewsky N et al. Probabilistic clustering of sequences: Inferring new bacterial regulons by comparative genomics. Proc Natl Acad Sci USA 2002; 99:7323-7328.
12. Pazos F, Valencia A. Similarity of phylogenetic trees as indicator of protein-protein interaction. Protein Eng 2001; 14:609-614.
13. Ramani AK, Marcotte EM. Exploiting the coevolution of interacting proteins to discover interaction specificity. J Mol Biol 2003; 327:273-284.
14. Goh CS, Bogan AA, Joachimiak M et al. Coevolution of proteins with their interaction partners. J Mol Biol 2000; 299:283-293.
15. Huynen MA, Snel B. Exploiting the variations in the genomic associations of genes to predict pathways and reconstruct their evolution. In: Koonin EV, ed. Frontiers in Computational Genomics. Wymondham: Caisters 2003:3:145-166.
16. Marcotte EM. Computational genetics: Finding protein function by nonhomology methods. Curr Opin Struct Biol 2000; 10:359-365.
17. Valencia A, Pazos F. Computational methods for the prediction of protein interactions. Curr Opin Struct Biol 2002; 12:368-373.
18. Huynen M, Snel B, Lathe W et al. Exploitation of gene context. Curr Opin Struct Biol 2000; 10:366-370.
19. Teichmann S, Babu M. Conservation of gene coregulation in prokaryotes and eukaryotes. Trends Biotechnol 2002; 20:407.
20. van Noort V, Snel B, Huynen MA. Predicting gene function by conserved coexpression. Trends Genet 2003; 19:238-242.
21. Stuart JM, Segal E, Koller D et al. A gene-coexpression network for global discovery of conserved genetic modules. Science 2003; 302:249-255.
22. Kelley BP, Sharan R, Karp RM et al. Conserved pathways within bacteria and yeast as revealed by global protein network alignment. Proc Natl Acad Sci USA 2003; 100:11394-11399.
23. Huynen MA, Snel B, van Noort V. Comparative genomics for reliable protein function prediction. Trends Genet 2004, (in press).
24. von Mering C, Krause R, Snel B et al. Comparative assessment of large-scale data sets of protein-protein interactions. Nature 2002; 417:399-403.
25. Jansen R, Yu H, Greenbaum D et al. A Bayesian networks approach for predicting protein-protein interactions from genomic data. Science 2003; 302:449-453.
26. Snel B, Bork P, Huynen M. Genome evolution. Gene fusion versus gene fission. Trends Genet 2000; 16:9-11.
27. Welch GR, Easterby JS. Metabolic channeling versus free diffusion: Transition-time analysis. Trends Biochem Sci 1994; 19:193-197.
28. Yanai I, Derti A, DeLisi C. Genes linked by fusion events are generally of the same functional category: A systematic analysis of 30 microbial genomes. Proc Natl Acad Sci USA 2001; 98:7940-7945.
29. Tsoka S, Ouzounis CA. Prediction of protein interactions: Metabolic enzymes are frequently involved in gene fusion. Nat Genet 2000; 26:141-142.
30. Mushegian AR, Koonin EV. Gene order is not conserved in bacterial evolution. Trends Genet 1996; 12:289-290.
31. Watanabe H, Mori H, Itoh T et al. Genome plasticity as a paradigm of eubacteria evolution. J Mol Evol 1997; 44(Suppl 1):S57-64.
32. Wolf YI, Rogozin IB, Kondrashov AS et al. Genome alignment, evolution of prokaryotic genome organization, and prediction of gene function using genomic context. Genome Res 2001; 11:356-372.
33. Overbeek RF, D'Souza M, Pusch M et al. Use of contiguity on the chromosome to infer functional coupling. In Silico Biol 1998; 2:93-108.
34. Moreno-Hagelsieb G, Trevino V, Perez-Rueda E et al. Transcription unit conservation in the three domains of life: A perspective from Escherichia coli. Trends Genet 2001; 17:175-177.
35. Overbeek R, Fonstein M, D'Souza M et al. The use of gene clusters to infer functional coupling. Proc Natl Acad Sci USA 1999; 96:2896-2901.
36. Blumenthal T, Evans D, Link CD et al. A global analysis of Caenorhabditis elegans operons. Nature 2002; 417:851-854.

37. Rodionov DA, Vitreschak AG, Mironov AA et al. Comparative genomics of thiamin biosynthesis in procaryotes. New genes and regulatory mechanisms. J Biol Chem 2002; 277:48949-48959.
38. van Nimwegen E, Zavolan M, Rajewsky N et al. Probabilistic clustering of sequences: Inferring new bacterial regulons by comparative genomics. Proc Natl Acad Sci USA 2002; 99:7323-7328.
39. Huynen MA, Snel B. Gene and context: Integrative approaches to genome analysis. Adv Protein Chem 2000; 54:345-379.
40. Gaasterland T, Ragan MA. Microbial genescapes: Phyletic and functional patterns of ORF distribution among prokaryotes. Microb Comp Genomics 1998; 3:199-217.
41. Liberles DA, Thoren A, von Heijne G et al. The use of phylogenetic profiles for gene prediction. Current Genomics 2002; 3:131-137.
42. von Mering C, Huynen M, Jaeggi D et al. STRING: A database of predicted functional associations between proteins. Nucleic Acids Res 2003; 31:258-261.
43. Koonin EV, Mushegian AR, Bork P. Nonorthologous gene displacement. Trends Genet 1996; 12:334-336.
44. Fryxell KJ. The coevolution of gene family trees. Trends Genet 1996; 12:364-369.
45. Hughes AL, Yeager M. Coevolution of the mammalian chemokines and their receptors. Immunogenetics 1999; 49:115-124.
46. Valencia A, Pazos F. Prediction of protein-protein interactions from evolutionary information. Methods Biochem Anal 2003; 44:411-426.
47. Pazos F, Helmer-Citterich M, Ausiello G et al. Correlated mutations contain information about protein-protein interaction. J Mol Biol 1997; 271:511-523.
48. Pazos F, Valencia A. In silico two-hybrid system for the selection of physically interacting protein pairs. Proteins 2002; 47:219-227.
49. Huynen M, Snel B, Lathe IIIrd W et al. Predicting protein function by genomic context: Quantitative evaluation and qualitative inferences. Genome Res 2000; 10:1204-1210.
50. von Mering C, Huynen MA, Jaeggi D et al. STRING - adatabase of predicted functional associations between proteins. Nucleic Acids Res 2003.
51. Yanai I, Mellor JC, DeLisi C. Identifying functional links between genes using conserved chromosomal proximity. Trends Genet 2002; 18:176-179.
52. Bairoch A, Apweiler R. The SWISS-PROT protein sequence database and its supplement TrEMBL in 2000. Nucleic Acids Res 2000; 28:45-48.
53. Kanehisa M, Goto S, Kawashima S et al. The KEGG databases at GenomeNet. Nucleic Acids Res 2002; 30:42-46.
54. Uetz P, Giot L, Cagney G et al. A comprehensive analysis of protein-protein interactions in Saccharomyces cerevisiae. Nature 2000; 403:623-627.
55. Ho Y, Gruhler A, Heilbut A et al. Systematic identification of protein complexes in Saccharomyces cerevisiae by mass spectrometry. Nature 2002; 415:180-183.
56. Eisenberg D, Marcotte EM, Xenarios I et al. Protein function in the post-genomic era. Nature 2000; 405:823-826.
57. Huynen MA, Snel B, von Mering C et al. Function prediction and protein networks. Curr Opin Cell Biol 2003; 15:191-198.
58. Morett E, Korbel JO, Rajan E et al. Systematic discovery of analogous enzymes in thiamin biosynthesis. Nat Biotechnol 2003; 21:790-795.
59. Tatusov RL, Natale DA, Garkavtsev IV et al. The COG database: New developments in phylogenetic classification of proteins from complete genomes. Nucleic Acids Res 2001; 29:22-28.
60. Lim K, Tempczyk A, Parsons JF et al. Crystal structure of YbaB from Haemophilus influenzae (HI0442), a protein of unknown function coexpressed with the recombinational DNA repair protein RecR. Proteins 2003; 50:375-379.
61. Yeung T, Mullin DA, Chen KS et al. Sequence and expression of the Escherichia coli recR locus. J Bacteriol 1990; 172:6042-6047.
62. Letunic I, Copley RR, Schmidt S et al. SMART 4.0: Towards genomic data integration. Nucleic Acids Res 2004; 32(Database issue):D142-144.
63. Hughes TR, Marton MJ, Jones AR et al. Functional discovery via a compendium of expression profiles. Cell 2000; 102:109-126.
64. Gonzalez-Parraga P, Hernandez JA, Arguelles JC. Role of antioxidant enzymatic defences against oxidative stress H(2)O(2) and the acquisition of oxidative tolerance in Candida albicans. Yeast 2003; 20:1161-1169.
65. Aldea M, Hernandez-Chico C, de la Campa AG et al. Identification, cloning, and expression of bolA, an ftsZ-dependent morphogene of Escherichia coli. J Bacteriol 1988; 170:5169-5176.
66. Santos JM, Freire P, Vicente M et al. The stationary-phase morphogene bolA from Escherichia coli is induced by stress during early stages of growth. Mol Microbiol 1999; 32:789-798.

67. Kim SH, Kim M, Lee JK et al. Identification and expression of uvi31+, a UV-inducible gene from Schizosaccharomyces pombe. Environ Mol Mutagen 1997; 30:72-81.
68. Ito T, Tashiro K, Muta S et al. Toward a protein-protein interaction map of the budding yeast: A comprehensive system to examine two-hybrid interactions in all possible combinations between the yeast proteins. Proc Natl Acad Sci USA 2000; 97:1143-1147.
69. Giot L, Bader JS, Brouwer C et al. A protein interaction map of Drosophila melanogaster. Science 2003; 302:1727-1736.
70. Kasai T, Inoue M, Koshiba S et al. Solution structure of a BolA-like protein from Mus musculus. Protein Sci 2004; 13:545-548.
71. Holm L, Sander C. Protein structure comparison by alignment of distance matrices. J Mol Biol 1993; 233:123-138.
72. Gabaldon T, Huynen MA. Reconstruction of the proto-mitochondrial metabolism. Science 2003; 301:609.
73. Sen GC, Lengyel P. The interferon system. A bird's eye view of its biochemistry. J Biol Chem 1992; 267:5017-5020.
74. Bisbal C, Martinand C, Silhol M et al. Cloning and characterization of a RNAse L inhibitor. A new component of the interferon-regulated 2-5A pathway. J Biol Chem 1995; 270:13308-13317.
75. Zimmerman C, Klein KC, Kiser PK et al. Identification of a host protein essential for assembly of immature HIV-1 capsids. Nature 2002; 415:88-92.
76. Letunic I, Goodstadt L, Dickens NJ et al. Recent improvements to the SMART domain-based sequence annotation resource. Nucleic Acids Res 2002; 30:242-244.
77. Stuart JM, Segal E, Koller D et al. A gene coexpression network for global discovery of conserved genetic modules. Science 2003; 302:249-255.
78. Ito T, Chiba T, Ozawa R et al. A comprehensive two-hybrid analysis to explore the yeast protein interactome. Proc Natl Acad Sci USA 2001; 98:4569-4574.
79. Krogan NJ, Peng WT, Cagney G et al. High-definition macromolecular composition of yeast RNA-processing complexes. Mol Cell 2004; 13:225-239.
80. Valasek L, Hasek J, Nielsen KH et al. Dual function of eIF3j/Hcr1p in processing 20 S prerRNA and translation initiation. J Biol Chem 2001; 276:43351-43360.
81. Huh WK, Falvo JV, Gerke LC et al. Global analysis of protein localization in budding yeast. Nature 2003; 425:686-691.
82. Kressler D, Linder P, de La Cruz J. Protein trans-acting factors involved in ribosome biogenesis in Saccharomyces cerevisiae. Mol Cell Biol 1999; 19:7897-7912.
83. Giaever G, Chu AM, Ni L et al. Functional profiling of the Saccharomyces cerevisiae genome. Nature 2002; 418:387-391.
84. Estevez AM, Haile S, Steinbuchel M et al. Effects of depletion and overexpression of the Trypanosoma brucei ribonuclease L inhibitor homologue. Mol Biochem Parasitol 2004; 133:137-141.
85. Fromme JC, Verdine GL. Structure of a trapped endonuclease III-DNA covalent intermediate. EMBO J 2003; 22:3461-3471.
86. Porello SL, Cannon MJ, David SS. A substrate recognition role for the [4Fe-4S]2+ cluster of the DNA repair glycosylase MutY. Biochemistry 1998; 37:6465-6475.
87. Weller GR, Kysela B, Roy R et al. Identification of a DNA nonhomologous end-joining complex in bacteria. Science 2002; 297:1686-1689.
88. Thomas G, Coutts G, Merrick M. The glnKamtB operon. A conserved gene pair in prokaryotes. Trends Genet 2000; 16:11-14.
89. Coutts G, Thomas G, Blakey D et al. Membrane sequestration of the signal transduction protein GlnK by the ammonium transporter AmtB. EMBO J 2002; 21:536-545.
90. Bobik TA, Rasche ME. Identification of the human methylmalonyl-CoA racemase gene based on the analysis of prokaryotic gene arrangements. Implications for decoding the human genome. J Biol Chem 2001; 276:37194-37198.
91. Horswill AR, Escalante-Semerena JC. In vitro conversion of propionate to pyruvate by Salmonella enterica enzymes: 2-methylcitrate dehydratase (PrpD) and aconitase Enzymes catalyze the conversion of 2-methylcitrate to 2-methylisocitrate. Biochemistry 2001; 40:4703-4713.
92. Daugherty M, Vonstein V, Overbeek R et al. Archaeal shikimate kinase, a new member of the GHMP-kinase family. J Bacteriol 2001; 183:292-300.
93. Graham DE, Graupner M, Xu H et al. Identification of coenzyme M biosynthetic 2-phosphosulfolactate phosphatase. A member of a new class of Mg(2+)-dependent acid phosphatases. Eur J Biochem 2001; 268:5176-5188.
94. Kurnasov O, Jablonski L, Polanuyer B et al. Aerobic tryptophan degradation pathway in bacteria: Novel kynurenine formamidase. FEMS Microbiol Lett 2003; 227:219-227.

95. Graham DE, Xu H, White RH. Methanococcus jannaschii uses a pyruvoyl-dependent arginine decarboxylase in polyamine biosynthesis. J Biol Chem 2002; 277:23500-23507.
96. Heath RJ, Rock CO. A triclosan-resistant bacterial enzyme. Nature 2000; 406:145-146.
97. Marrakchi H, Choi KH, Rock CO. A new mechanism for anaerobic unsaturated fatty acid formation in Streptococcus pneumoniae. J Biol Chem 2002; 277:44809-44816.
98. Bishop AC, Xu J, Johnson RC et al. Identification of the tRNA-dihydrouridine synthase family. J Biol Chem 2002; 277:25090-25095.
99. Huynen MA, Snel B, Bork P et al. The phylogenetic distribution of frataxin indicates a role in iron-sulfur cluster protein assembly. Hum Mol Genet 2001; 10:2463-2468.
100. Muhlenhoff U, Richhardt N, Ristow M et al. The yeast frataxin homolog Yfh1p plays a specific role in the maturation of cellular Fe/S proteins. Hum Mol Genet 2002; 11:2025-2036.
101. Luttgen H, Rohdich F, Herz S et al. Biosynthesis of terpenoids: YchB protein of Escherichia coli phosphorylates the 2-hydroxy group of 4-diphosphocytidyl-2C-methyl-D-erythritol. Proc Natl Acad Sci USA 2000; 97:1062-1067.
102. Karzai AW, Susskind MM, Sauer RT. SmpB, a unique RNA-binding protein essential for the peptide-tagging activity of SsrA (tmRNA). EMBO J 1999; 18:3793-3799.
103. Myllykallio H, Lipowski G, Leduc D et al. An alternative flavin-dependent mechanism for thymidylate synthesis. Science 2002; 297:105-107.
104. Rouhier N, Gelhaye E, Sautiere PE et al. Isolation and characterization of a new peroxiredoxin from poplar sieve tubes that uses either glutaredoxin or thioredoxin as a proton donor. Plant Physiol 2001; 127:1299-1309.
105. Herz S, Wungsintaweekul J, Schuhr CA et al. Biosynthesis of terpenoids: YgbB protein converts 4-diphosphocytidyl-2C-methyl-D-erythritol 2-phosphate to 2C-methyl-D-erythritol 2,4-cyclodiphosphate. Proc Natl Acad Sci USA 2000; 97:2486-2490.
106. Kryukov GV, Kumar RA, Koc A et al. Selenoprotein R is a zinc-containing stereo-specific methionine sulfoxide reductase. Proc Natl Acad Sci USA 2002; 99:4245-4250.
107. Campbell JW, Cronan Jr JE. The enigmatic Escherichia coli fadE gene is yafH. J Bacteriol 2002; 184:3759-3764.
108. Sadovskaya NS, Laikova ON, Mironov AA et al. Study of regulation of long-Chain Fatty acid metabolism using computer analysis of complete bacterial genomes. Mol Biol 2001; 35:862-866.
109. Rodionov DA, Mironov AA, Rakhmaninova AB et al. Transcriptional regulation of transport and utilization systems for hexuronides, hexuronates and hexonates in gamma purple bacteria. Mol Microbiol 2000; 38:673-683.
110. Hugouvieux-Cotte-Pattat N, Blot N, Reverchon S. Identification of TogMNAB, an ABC transporter which mediates the uptake of pectic oligomers in Erwinia chrysanthemi 3937. Mol Microbiol 2001; 41:1113-1123.
111. Zhang Z, Feige JN, Chang AB et al. A transporter of Escherichia coli specific for L- and D-methionine is the prototype for a new family within the ABC superfamily. Arch Microbiol 2003; 180:88-100.
112. Uetz P, Giot L, Cagney G et al. A comprehensive analysis of protein-protein interactions in Saccharomyces cerevisiae. Nature 2000; 403:623-627.
113. Gabaldon T, Huynen MA. Prediction of protein function and pathways in the genome era. Cell Mol Life Sci 2004; 61:930-944.

CHAPTER 2

Clues from Three-Dimensional Structure Analysis and Molecular Modelling: New Insights into Cytochrome P450 Mechanisms and Functions

Karin Schleinkofer and Thomas Dandekar*

Abstract

Cytochrome P450 is a focus of attention as it comprises one of the largest superfamilies of enzyme proteins. Metabolization of many drugs is affected by cytochrome P450. It is an attractive drug target, e.g., cytochrome P450s of *Mycobacterium tuberculosis* are promising targets in the fight against tuberculosis. The structure provides new insights for investigation of structure/mechanism of cytochrome P450, and for rational design of inhibitor molecules. We will illustrate how biocomputing and bioinformatical techniques reveal details, functions and further secrets of this exciting molecule. Molecular modelling along with site-directed mutagenesis of P450 2B1 elucidated the molecular determinants of substrate specifity. Regioselectivity of progesterone hydroxylation by cytochrome P450 2B1 was reengineered based on the X-ray structure of cytochrome 2C5. Docking approaches rationalized the regioselectivity of the reengineered cytochrome P450 2B1. Furthermore, by methods of molecular dynamic simulations, routes were identified by which substrates may enter into and products exit from the active site of cytochrome P450.

Introduction

Cytochrome P450 enzymes[1-5] form an ubiquitous heme protein monooxygenase family (EC: 1.14.14.1). They play an important role in the synthesis and degradation of many physiologically important compounds such as steroid hormones, cholesterol, bile acids and in the detoxification of xenobiotics in many species of microorganisms, plants and animals.

P450 are of great medical relevance: Mutations in P450 genes are triggers of human diseases such as primary congenital glaucoma and there are evidences for associations between cytochrome P450 enzyme-polymorphism and cancer. Some P450 enzymes are able to activate procarcinogens to genotoxic intermediates. They play a major role in drug-metabolism, for example the P450 3A family of enzymes are able to metabolize the majority of commercially available drugs such as Codeine (narcotic), Diazepam (Valium), Erythromycin (antibiotic). Drug metabolism polymorphism or interactions with other drugs can cause severe sideeffects in patients.

*Corresponding Author: Thomas Dandekar—Dept. Bioinformatik, Biozentrum, Am Hubland, D-97074 Universitaet Wuerzburg, F.R. Germany.
Email: dandekar@biozentrum.uni-wuerzburg.de

Discovering Biomolecular Mechanisms with Computational Biology, edited by Frank Eisenhaber.

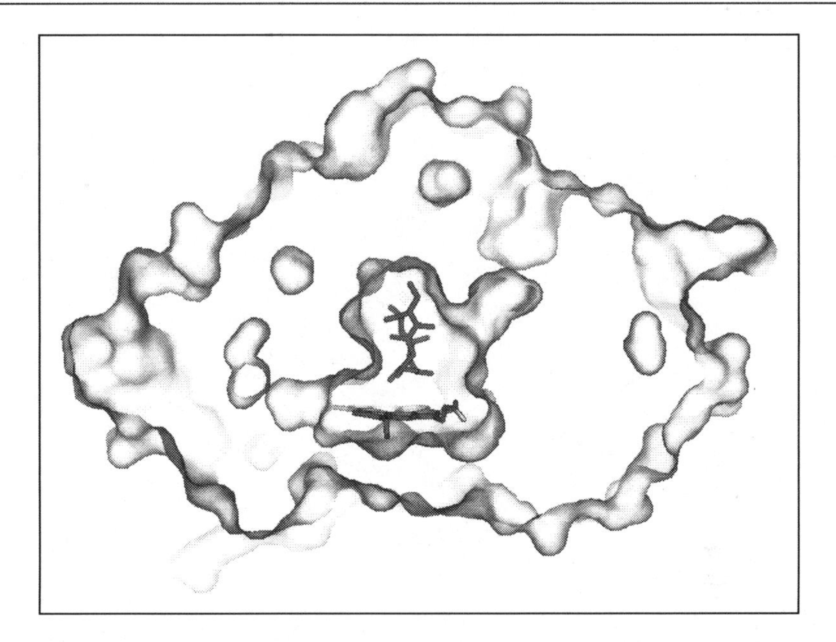

Figure 1. The active site molecular surface and the outer molecular surface of cytochrome P450eryF (CYP107A1) are not connected with each other. A substrate molecule, 6-deoxyerythonolide, is shown in the active site above the heme. The molecular surface was computed with the PyMOL program.[53]

P450 are heme-thiolate containing proteins where the ligand of the heme iron is delivered by a cysteine residue in a highly conserved region of the enzyme. The active site is buried at the center of the enzyme (Fig. 1). They are named P450 for the absorption band at 450 nm of their carbon-monoxide-bound form. The reactions carried out by cytochrome P450 molecules are very diverse and include hydroxylation, N-, O- and S-dealkylation and oxidation of heteroatoms.

According to their sequence similarity P450 enzymes are subdivided into families (sequence identity greater than 40%) and subfamilies (sequence identity greater than 55%). In humans 57 CYP genes are sequenced (and 58 pseudogenes) which are subdivided into 18 families and 43 subfamilies.

In prokaryotes P450 are soluble proteins whereas in eukaryotes P450 are usually membrane-associated within the inner mitochondrial membrane or endoplasmic reticulum.

Because of their physiological importance and medical relevance the P450 enzymes are an emerging field of research. Major unresolved issues are structurefunction relationships such as the understanding of substrate specificity, the catalytic mechanism of multi-step reactions, the dynamical properties that allow substrates to enter the active site and products to leave the active site or the identification of essential determinants of drug metabolism or tolerance. In the following paragraphs methods of computational biology are presented which aid our understanding of this interesting enzyme. However the presented methods are applicable to a variety of biomolecules.

Modelling

The gap between the high number of known protein sequences and the only limited available 3-dimensional protein-structures is increasing rapidly. Molecular modelling techniques are valuable tools to fill this gap.[6,7] In the field of cytochrome P450 research this technique is of high interest. Up to now more than 3700 cytochrom P450 (different named) sequences of different species are known, the determination of all these protein structures is a tedious work, because crystallization of some P450 enzymes, especially of the membrane-associated ones is

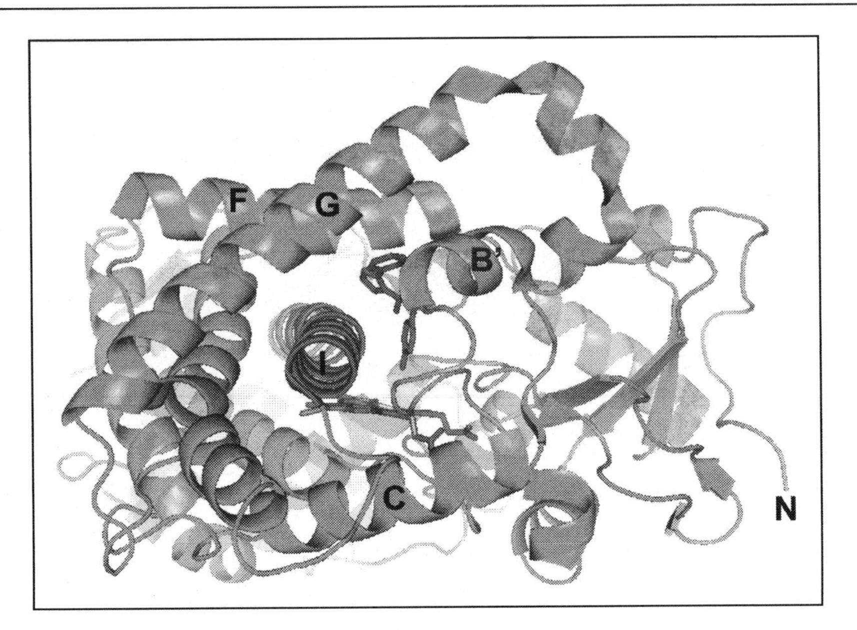

Figure 2. Structure of cytochrome P450 2C5 bound to a substrate. A substrate molecule, 4-methyl-*N*-methyl-*N*-(2-phenyl-2*H*-pyrazol-3-yl)benzenesulfonamide, is shown in the active site above the heme. The principal helices and the NH_2-terminal are labelled. P450 2C5 is membrane-associated by its N-terminal tip.

difficult. So far more than 120 P450s structures are in the brookhaven protein database. Among them two mammalian P450 enzymes (2C5 and 2B4), the first two structures of membrane-associated P450s[8,9] which were solved recently (Fig. 2). The overall fold of P450s is conserved despite their low sequence identity (as low as 10%).

Homology Modelling

Although the amino acid sequence must finally determine a protein's three dimensional structure and despite of intensive research the development of an algorithm to determine the accurate 3D-structure from amino acid sequence has yet not been achieved.

The most promising approach is the modelling based on structures of homologue proteins. The progress within the field of protein structure predicition by NMR or X-ray crystallography enhances the probability to find a homologue protein of which the 3-dimensional structure is known.

Experimentally determined protein and DNA/RNA structures can be found in the brookhaven protein databank (http://www.pdb.org). To compute a good model of the unknown structure the protein homologue should have a sequence identity of more than 30%. The prerequiste of a good model is the generation of an alignment between sequence of the experimentally determined template structure and the protein which has to be modeled. To generate a satisfactory alignment, experimental data should also be taken into account such as site-directed mutagenesis data of residues known to be involved in substrate binding, binding to redox partners; anibody recognition sequences which indicate whether certain residues are located on the surface of a protein.

The prediction of the secondary structure elements is usually reasonable accurate (~75%) for a three state prediction using sequence alignments whereas the prediction within the loop-region and of dynamic sidechains could be problematic. In order to refine the structure the model should be energy minimized using molecular mechanics.

There are comparative modelling servers on the web such as SWISS-MODEL,[10] CPHmodels.[11] Software packages for detailed protein structure analysis include Modeller,[12] WHATIF,[13] Insight[14] and Sybyl.[15]

The automated SWISS-MODEL software works in several steps as follows:

- superposition of homologue 3D-structures
- generation of a multiple alignment
- calculation of peptidchain form the averaged coordinates
- reconstruction of loops based on a coordinate library
- addition and correction of sidechains
- validation of the structure (checking of stereochemical quality)

Substrate specificity and type of reactions catalyzed are governed by less conserved regions of P450s and are therefore not well understood. In this context the prediction of the regioselectivity of an enzymatic reaction is of particular interest. In a regioselective reaction one possible product out of two or more is formed preferentially (it is often the case that addition and elimination reactions may, in principle, proceed to more than one product—these are often isomers of each other).

In the following approach based on the 3D model of P450 2B1 it was shown that active site residues are responsible for regioselectivity.[16] P450 2B1 belongs to the 2B subfamily comprising enzymes with a broad range of substrates, including drugs, environmental carcinogens and steroids. A model of 2B1 was built using the X-ray structure of P450 2C5 as a template. The location of the active site residues within the 3D P450 model of 2B1 can be visualized. The active site residues of 2B1 could be deduced from the 2C5 structure and were verified experimentally by site-directed mutagenesis. 2B1 has progesterone 16α-hydroxylase activity whereas 2C5 has progesterone 21-hydroxylation activity. By replacing seven active site residues of 2B1 by the corresponding active site residues of 2C5 a novel progesterone 21-hydroxylation activity was confered to 2B1. The mutated 2B1 showed 80% regioselectivity for progesterone 21-hydroxylation.

Threading

In case that there is no homologue protein or the sequence identity is very low (e.g., a novel cytochrome from Archaebacteria) the threading approach is a good alternative.[17-19] The starting point of a threading approach are protein folds. It is known that proteins having no sequence similarity can have similar 3D-structures. Examples are actin and hexokinase which exhibit the same Ribonuclease H-like folding topology,[20] despite having different sequences. In the course of evolution only a limited number of protein folds emerged (~1000).[21] The sequence for which a prediction is required is threaded onto all known protein folds. As current databases have already covered a large part of protein folds generally used in nature, this approach is successful. Useful Webservers include: genThreader[22,23] and 3D-PSSM.[24]

Ab Initio Modelling

Even novel protein folds could be predicted by ab initio modelling. Ab initio structure prediction requires only the sequence of a protein to generate a 3-dimensional model. This approach is computationally demanding, there are several algorithms which for example rely on physicochemical energetics, or on methods that use predicted secondary structure in combination with distance constraints.[25-27] The catalytic arrangement of the cytochrome P450 center provides useful distance constraints[28] and can be confirmed by conserved residues in these positions. This technique could be used to improve ab initio models of cytochromes P450 with completely novel folds.

Assessing Tertiary Structure Prediction

No matter which method was used to establish a structural model, it is essential to assess the validity of the generated model by checking how well the new model conforms to protein stereochemical quality. There is some agreement about which measurements are good indicators of stereochemical quality; these include planarity; chirality; phi/psi preferences; chi angles; nonbonded contact distances; unsatisfied donors and acceptors. The ProCheck package comprises a number of complementary procedures for evaluating protein structures and identifies regions of the modeled protein which may require further refinement. The following webservers are very useful: Whatcheck,[29] ProCheck.[30] These tools can be used to check the model quality in the study on cytochrome regioselectivity as residue positions are critical here.

Modelling of Protein-Ligand Complexes

An important and useful area of molecular modelling is the modelling of protein-ligand complexes. Comparative methods and docking approaches can be applied. Protein docking methods help understanding the mecanism of molecular interaction and are also useful in the development of novel pharmaceutical agents because it helps to screen out unfavorable ligands at an early stage.

A docking procedure[31-34] can be subdivided into three processes: identification of the binding site, sampling of possible ligand orientations and positions in the binding site and scoring of the possible sampled solutions. The rigid-body model is historically the first docking approach where the flexibility of the interaction partners is not considered. However most of the current rigid-body methods could address ligand flexibilty by accounting an ensemble of ligand conformations or allowing some intermolecular interactions.

Software programs carrying out docking calculations are for example DOCK,[35] GRID,[36,37] AutoDock,[38] FTDOCK.[39] The DOCK program suite is one of the oldest and best known ligand-protein docking programs. In newer versions of DOCK the ligand flexibility is incooperated.

In a first step a 'negative image' of the binding site is constructed by overlapping spheres of varying radii. In a second step the ligand atoms are matched to these sphere centres to position the molecule within the binding site. GRID allows the identification of ligand binding sites of a protein. The protein is put into a 3-dimensional grid, at each point of the grid the molecular mechanics interaction energy between the protein and a series of probe molecules is computed. Probe molecules are a series of chemical functional groups such as phosphate, methyl, hydroxyl, carboxyl. AutoDock identifies the binding site by a genetic algorithm search. Estimated binding free energies can also be computed.

Many computational methods used for modelling of protein-ligand complexes can also be applied to model protein-protein-complexes. For example the DOCK program has been used to model protein-protein complexes. To accelerate the search for the best geometric fit between two proteins, Fourier transform methods are applied, a shape recognition algorithm in which molecules are discretised onto grid. All possible translations of the molecules are scanned by superposition of the grid points. The program FTDock is based on Fourier transform procedure and a method to determine electrostatic complemenarity.

Docking of inhibitors or substrate molecules into the active site of P450 aid understanding the key enzyme-substrate-interactions as well as the role of particular residues in catalysis. Steric considerations as well as orientating the site of metabolism toward heme and ferryl oxygen has to be taken into account.

In the following we describe one particular drug target and how docking methods can help in the development of new therapeuticals. Cytochrom P450 (CYP121) of Mycobacterium tuberculosis (Mtb) is a promising drug target to combat the multidrug-resistant strains of Mycobacterium tuberculosis.[40] The genome of Mycobacterium tuberculosis reveals an exceptional high number of 20 different encoded P450—the highest number in any bacterium. Several azole drugs which are known inhibitors of cytochrome P450 have been shown to have potent antimycobacterial activity, especially high affinity for CYP121. But many azol anti-fungal drugs show cross-reactivity with human P450 isoforms. Docking studies should help to rationalize the key determinant that dictate tight CYP121-drug interactions in order to design novel azole-based drugs that have high selectivity for CYP121. First docking studies could rationalize binding of certain azole drugs, e.g., miconazole whereas bulkier azole drugs failed to dock into the active site.

Molecular Dynamic Simulation

Molecular dynamic simulations[41,42] have become a standard tool for the investigation of biomolecules. Simulations give insights into the natural dynamics of biomolecules and thereby aid our understanding of biochemical processes. Especially in protein-ligand interaction the flexibility of the interaction partners is of great importance. In order to fit in the active site of the protein the ligand has to adopt a certain 3-dimensional conformation. However, a protein is despite being a rigid entity not completely stiff. Especially the positions of the amino acid side chains can be very flexible and change their positions while the protein is „at work": For lots of enzymatic processes and binding of ligands this flexibility is essential for a protein's specific function. These subtle movements are the basis for all metabolic processes and thus are a key process of what life, a living protein "breathes". Even protein domains can change their position relatively to each other. The experimental approaches to study biomolecular dynamics are still limited. With continuing advances in the speed of computers and the methodology of molecular dynamic simulations the time scales that are becoming available are making it possible to study phenomena of biological interest in real time.

For the optimization of crystal structures and NMR structures molecular dynamic simulations are applied as a standard tool. The principle of the molecular dynamic simulation is to record the movement of atoms under the influence of a selected force field (OPLS,[43] CHARMM,[44] GROMOS96[45] or AMBER[46]). It is assumed that the movement of the atoms follows classical mechanics. The theoretical basis of a molecular dynamic simulation is Newton's equation of motion ($F_i(t) = m_i \cdot a_i(t)$, where F is the force acting on atom i at time t, m is the mass of atom i, a the acceleration acting on atom i at time t), which is solved numerically for each atom. Forces between the atoms are neglected during the molecular dynamic simulation. The initial point of a simulation is a starting structure, which is in most cases an experimentally determined structure. To simulate the dynamics of a whole protein it is solvated with water molecules. At the start of a simulation a velocity is assigned to each atom, which corresponds to the selected temperature of the simulation. Newton's equation of motion is solved for each atom taking these initial values into account and the computed coordinates are saved periodically.

In the P450 research molecular dynamic simulations were applied to address the question how substrates enter or exit the isolated cavity of the enzyme. To allow substrate access and product exit the enzyme must undergo structural motions. The understanding of these motions would also help to explain why P450s have such a broad diversity of substrates and such a wide variation in degree of specificity. Furthermore the enzyme kinetic of P450 was shown to be influenced by protein dynamics. In the following we describe two molecular dynamics simulation methods elucidating potential ligand exit pathways in P450s. Steered molecular dynamic simulations (SMD)[47-49] of testosterone exit suggested a functional role for the residues in the N-terminal portion of the cytochrome P450 2B1 I helix.[50] These data are confirmed by site-directed mutagenesis data within the I-helix as these alter the enzymatic activity of the enzyme. SMD is an extended MD simulation method mimicking the principle of the atomic force microscopy (AFM). In SMD simulations, time-dependent external forces are applied to the ligand to facilitate its dissociation from the protein by movement along a trajectory. Because position restraints are removed from the entire protein-ligand complex, the SMD simulation allows the protein to be repositioned in response to the accelerated dissociation process of the ligand.

By random expulsion molecular dynamic simulations (REMD)[51] the substrate exit of different P450 enzymes could be identified. An investigation by REMD of mammalian cytochrome P450 showed that the substrate egress is different from that of soluble, bacterial P450.[52] In REMD the probability of spontaneous substrate exit in the time range amenable

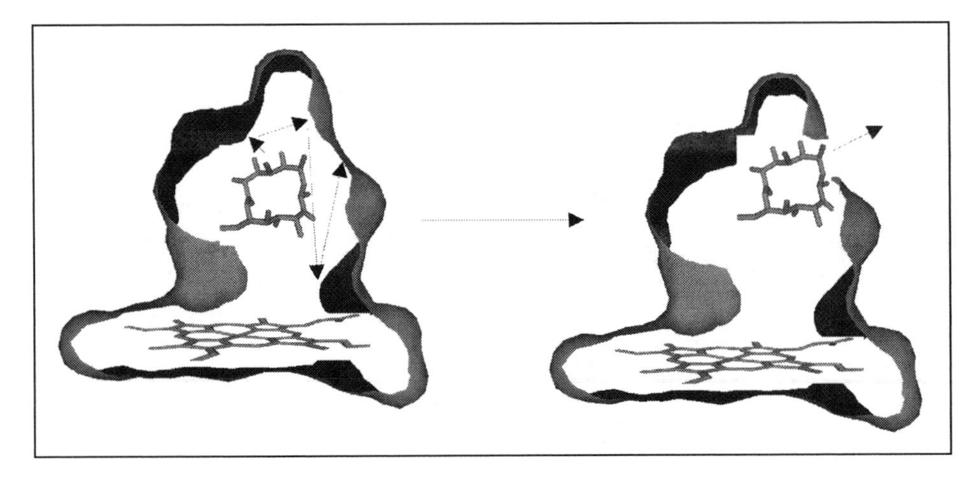

Figure 3. In REMD, routes by which a ligand can exit from an interior cavity in a protein are identified (here the active site of a cytochrome P450 with a substrate molecule above the heme is shown). An artificial randomly orientated force is applied to the center of mass of the ligand.

to molecular dynamic simulation is enhanced by an artificial force with random direction imposed upon the substrate in addition to the standard force field (Fig. 3). The direction of the additional force acting on the center of mass of the ligand is chosen randomly. The direction of the force is kept for a chosen number of time steps, N. During this time period, a specified distance r_{min}, should be covered by the substrate molecule: the substrate is required to travel at an average threshold velocity v during the time period $N \times \Delta t$, where Δt is the time step of the molecular dynamic simulation. If the substrate encounters relatively rigid parts of the cavity its average velocity will fall below the preset threshold. In this case a new direction is chosen randomly and maintained, as long as the substrate moves in the new direction with an average velocity larger than the preset value. In this way, the substrate probes different regions of the protein during the simulation until it exits.

References

1. Schuler MA, Werck-Reichhart D. Functional genomics of P450s. Annu Rev Plant Biol 2003; 54:629-667.
2. Werck-Reichhart D, Feyereisen R. Cytochromes P450: A success story. Genome Biol 2000; 1(6):REVIEWS3003.
3. Graham-Lorence S, Peterson JA. P450s: Structural similarities and functional differences. FASEB J 1996; 10(2):206-214.
4. Mueller EJ, Loida PJ, Sligar SG. In: Ortiz de Montellano PR, ed. Cytochrome P450: Structure, Mechanism, and Biochemistry. New York and London: Plenum Press, 2003:83-124.
5. David Nelson's homepage. http://drnelson.utmem.edu/CytochromeP450.html.
6. Russell RB, Sternberg MJ. Structure prediction. How good are we? Curr Biol 1995; 5(5):488-490.
7. Sali A, Overington JP, Johnson MS et al. From comparisons of protein sequences and structures to protein modelling and design. Trends Biochem Sci 1990; 15(6):235-240.
8. Wester MR, Johnson EF, Marques-Soares C et al. Structure of a substrate complex of mammalian cytochrome P450 2C5 at 2.3 A resolution: Evidence for multiple substrate binding modes. Biochemistry 2003; 42(21):6370-6379.
9. Scott EE, White MA, He YA et al. Structure of mammalian cytochrome P450 2B4 complexed with 4-(4-chlorophenyl)imidazole at 1.9-A resolution: Insight into the range of P450 conformations and the coordination of redox partner binding. J Biol Chem 2004; 279(26):27294-27301.
10. Guex N, Peitsch MC. SWISS-MODEL and the swiss-Pdbviewer: An environment for comparative protein modeling. Electrophoresis 1997; 18(15):2714-2723.
11. Lund O, Nielsen M, Lundegaard P et al. CPH models 2.0:X3M a computer program to extract 3D models. Abstract at the CASP5 conference 2002; A102.

12. Marti-Renom MA, Stuart AC, Fiser A et al. Comparative protein structure modeling of genes and genomes. Annu Rev Biophys Biomol Struct 2000; 29:291-325.
13. Rodriguez R, Chinea G, Lopez N et al. Homology modeling, model and software evaluation: Three related resources. Bioinformatics 1998; 14(6):523-528.
14. InsightII. San Diego, CA: Accelerys.
15. SYBYL 6.5. South Hanley Rd., St. Louis, Missouri, USA: Tripos Inc., 1699:63144.
16. Kumar S, Scott EE, Liu H et al. A rational approach to Reengineer cytochrome P450 2B1 regioselectivity based on the crystal structure of cytochrome P450 2C5. J Biol Chem 2003; 278(19):17178-17184.
17. Jones DT, Thornton JM. Potential energy functions for threading. Curr Opin Struct Biol 1996; 6(2):210-216.
18. Jones DT. Progress in protein structure prediction. Curr Opin Struct Biol 1997; 7(3):377-387.
19. Sippl MJ, Flockner H. Threading thrills and threats. Structure 1996; 4(1):15-19.
20. Murzin AG, Brenner SE, Hubbard T et al. SCOP: A structural classification of proteins database for the investigation of sequences and structures. J Mol Biol 1995; 247:536-540.
21. Koonin EV, Wolf YI, Karev GP. The structure of the protein universe and genome evolution. Nature 2002; 420(6912):218-223.
22. Jones DT. GenTHREADER: An efficient and reliable protein fold recognition method for genomic sequences. J Mol Biol 1999; 287(4):797-815.
23. McGuffin LJ, Jones DT. Improvement of the GenTHREADER method for genomic fold recognition. Bioinformatics 2003; 19(7):874-881.
24. Kelley LA, MacCallum RM, Sternberg MJ. Enhanced genome annotation using structural profiles in the program 3D-PSSM. J Mol Biol 2000; 299(2):499-520.
25. Aloy P, Stark A, Hadley C et al. Predictions without templates: New folds, secondary structure, and contacts in CASP5. Proteins 2003; 53(Suppl 6):436-456.
26. Kinch LN, Wrabl JO, Krishna SS et al. CASP5 assessment of fold recognition target predictions. Proteins 2003; 53(Suppl 6):395-409.
27. Tramontano A, Morea V. Assessment of homology-based predictions in CASP5. Proteins 2003; 53(Suppl 6):352-368.
28. Dandekar T, Argos P. Applying experimental data to protein fold prediction with the genetic algorithm. Protein Eng 1997; 10(8):877-893.
29. Hooft RW, Vriend G, Sander C et al. Errors in protein structures. Nature 1996; 381(6580):272.
30. Laskowski RA, MacArthur MW, Moss DS et al. PROCHECK: A program to check the stereochemical quality of protein structures. J Appl Cryst 1993; 26:283-291.
31. Halperin I, Ma B, Wolfson H et al. Principles of docking: An overview of search algorithms and a guide to scoring functions. Proteins 2002; 47(4):409-443.
32. Smith GR, Sternberg MJ. Prediction of protein-protein interactions by docking methods. Curr Opin Struct Biol 2002; 12(1):28-35.
33. Abagyan R, Totrov M. High-throughput docking for lead generation. Curr Opin Chem Biol 2001; 5(4):375-382.
34. Shoichet BK, McGovern SL, Wei B et al. Lead discovery using molecular docking. Curr Opin Chem Biol 2002; 6(4):439-446.
35. Ewing TJ, Makino S, Skillman AG. DOCK 4.0: Search strategies for automated molecular docking of flexible molecule databases. J Comput Aided Mol 2001; 15(5):411-428.
36. Goodford PJ. A computational procedure for determining energetically favorable binding sites on biologically important macromolecules. J Med Chem 1985; 28(7):849-857.
37. Wade RC, Goodford PJ. Further development of hydrogen bond functions for use in determining energetically favorable binding sites on molecules of known structure. 2. Ligand probe groups with the ability to form more than two hydrogen bonds. J Med Chem 1993; 36(1):148-156.
38. Morris GM, Goodsell DS, Halliday RS et al. Automated docking using a lamarckian genetic algorithm and and empirical binding free energy function. J Comput Chem 1998; 19:1639-1662.
39. Gabb HA, Jackson RM, Sternberg MJ. Modelling protein docking using shape complementarity, electrostatics and biochemical information. J Mol Biol 1997; 272(1):106-120.
40. Munro AW, McLean KJ, Marshall KR et al. Cytochromes P450: Novel drug targets in the war against multidrug-resistant Mycobacterium tuberculosis. Biochem Soc Trans 2003; 31(Pt 3):625-630.
41. Hansson T, Oostenbrink C, van Gunsteren W. Molecular dynamics simulations. Curr Opin Struct Biol 2002; 12(2):190-196.
42. Karplus M, McCammon JA. Molecular dynamics simulations of biomolecules. Nat Struct Biol 2002; 9(9):646-652.
43. Jorgensen W, Tirado-Rives J. The OPLS potential functions for proteins. Energy minimizations for crystals of cyclic peptides and crambin. J Am Chem Soc 1988; 110:1657-1666.

44. Brooks B, Bruccoleri R, Olafson B et al. CHARMM: A program for macromolecular energy, minimization, and dynamics calculations. Comp Chem 1983; 4:187-217.
45. Schuler LD, Daura X, van Gunsteren WF. An improved GROMOS96 force field for aliphatic hydrocarbons in the condensed phase. J Comp Chem 2001; 22:1205.
46. Cornell WD, Cieplak P, Payly CI et al. A second generation force field for the simulation of proteins, nucleic acids and organic molecules. J Am Chem Soc 1995; 117:5179-5197.
47. Grubmuller H, Heymann B, Tavan P. Ligand binding: Molecular mechanics calculation of the streptavidin-biotin rupture force. Science 1996; 271(5251):997-999.
48. Isralewitz B, Gao M, Schulten K. Steered molecular dynamics and mechanical functions of proteins. Curr Opin Struct Biol 2001; 11(2):224-230.
49. Izrailev S, Stepaniants S, Balsera M et al. Molecular dynamics study of unbinding of the avidin-biotin complex. Biophys J 1997; 72(4):1568-1581.
50. Scott EE, Liu H, Qun He Y et al. Mutagenesis and molecular dynamics suggest structural and functional roles for residues in the N-terminal portion of the cytochrome P450 2B1 I helix. Arch Biochem Biophys 2004; 423(2):266-276.
51. Ludemann SK, Lounnas V, Wade RC. How do substrates enter and products exit the buried active site of cytochrome P450cam? 1. Random expulsion molecular dynamics investigation of ligand access channels and mechanisms. J Mol Biol 2000; 303(5):797-811.
52. Wade RC, Winn PJ, Schlichting I et al. A survey of active site access channels in cytochromes P450. J Inorg Biochem 2004; 98(7):1175-1182.
53. DeLano WL. The PyMOL molecular graphics system on world wide web http://www.pymol.org.; 2002.

CHAPTER 3

Prediction of Protein Function:

Two Basic Concepts and One Practical Recipe

Frank Eisenhaber*

Abstract

The analysis of uncharacterized biomolecular sequences obtained as a result of genetic screens, expression profile studies, etc. is a standard task in a life science research environment. The understanding of protein function is typically the main difficulty. This chapter intends to give practical advise to students and researchers that have only introductory knowledge in the field of protein sequence analysis.

Applicable theoretical approaches range from (1) textual analyses, interpretation in terms of patterns of physical properties of amino acid side chains and (2) the extrapolation of empirically established relationships between local sequence motifs with known structural and functional properties to the collection of sequence segment families with sequence distance metrics and protein function derivation with annotation transfer (concept of homologous families). Here, the impact of different techniques for the biological interpretation of targets is discussed from the practitioner's point of view and illustrated with examples from recent research reports. Although sequence similarity searching techniques are the most powerful instruments for the analysis of high-complexity regions, other techniques can supply important additional evaluations including the assessment of applicability of the sequence homology concept for the given target segment.

Introduction

The genome has become the integrating principle for the various fields of biology and the clarification of pathways that lead to the realization of genome information into phenotypes under varying environmental conditions has become the central task for life sciences. As a first step, it is critical to understand the function of genes at least in qualitative terms; i.e., to name the molecular function of encoded proteins and to uncover the topology of interactions of networks involving them. Given that, currently, the molecular function of at least two thirds of all genes in completely sequenced eukaryote genomes remains more or less clouded, this would represent a dramatic progress. At the same time, it should be noted that real theoretical predictability of biological systems above the level of educated guesses (for example, for drug engineering) typically requires quantitative characterization of gene and protein activity and modeling of biological networks, which will be, in most cases, not a matter of the coming handful of years. Possibly, this is even an optimistic assessment.

With the central role of the genome in the functioning of biological systems, it is not surprising that experimental screens for genes relevant for the processes investigated are a standard approach in today's experimental biology; for example, expression profiling with DNA

*Frank Eisenhaber—Research Institute of Molecular Pathology, Dr. Bohr-Gasse 7, A-1030 Vienna, Republic Austria. Email: Frank.Eisenhaber@imp.univie.ac.at

Discovering Biomolecular Mechanisms with Computational Biology, edited by Frank Eisenhaber.

microarrays, yeast two hybrid screens, etc. If the biological phenomenon has not been well described in already published research, the screens lead typically to sequence tags of yet uncharacterized genes. Their sequence information has then to be interpreted in functional terms within the given physiological context. Stereotypically, the sequence is submitted to a similarity search in sequence databases. As a rule, the amount of insight produced by such a direct approach is indirectly proportional to the novelty of the gene target. In this tractate, we want to discuss the few fundamental principles that underlie state-of-the-art protein sequence analysis approaches. Then, we propose a general recipe for the practitioner who looks for research hints in his target sequences. We will give interpretation guides for sequence analytic findings and emphasize limitations where appropriate.

The Beginning: Deriving the Protein Sequence and the Definition of Protein Function

Typically, the starting point is a partial nucleic acid sequence representing a piece of mRNA. Whereas the experimental extension of the sequence to a full transcript was mandatory before the era of large-scale sequencing, this step can often be avoided now. In this case, it is necessary to find (1) a longer expressed sequence tag (EST), (2) a cluster of ESTs with a consensus sequence or, luckily, (3) a complete cDNA in the databases that obviously contains the reliably sequenced segment of the partial sequence obtained in the screen. The completeness of the putative transcript sequence can be investigated by mapping relevant ESTs onto the genome sequence. Especially in the case of incomplete transcripts involving only 3' untranslated regions, searching for the closest predicted gene upstream in the genome might yield the desired gene.[1,2] Searches for ESTs that bridge the distance between the detected gene and the mapped site are a possible reliability check and can also discriminate cases of alternative splicing. Further, the possibility of stumbling onto a pseudogene must be ruled out.[3,4]

Whereas all the steps leading to the protein sequences possibly encoded in the given transcript (in this essay, we do not consider untranslated RNAs) are sometimes complicated by sequencing errors (frameshifts, single point exchanges, genome fusion errors) but, in most cases, are just a technical exercise, the insufficient understanding of biological function for proteins known only as conceptual translations has become the major bottleneck in sequence data interpretation.

A few words on protein function: Protein function requires a hierarchical concept for the description of its many aspects that reflects the complexity of living systems.[5] The protein's function at the molecular level is rather a list of potential capabilities determined by its primary and tertiary structure. *Molecular function* description includes qualitative and quantitative aspects of diffusion properties in solution and membrane environments, conformational flexibility, allosteric conformational changes, possible ligand-binding (or catalytic) activities and ability for posttranslational modifications. Depending on cellular context (subcellular localization), different features of the molecular function may become important. A set of many cooperating proteins is responsible for a *cellular function* (metabolic pathway, signal transduction cascade, cytoskeletal complex, etc.). Since gene expression is regulated in a time- and tissue-dependent manner, regulatory sequences in the genomic environment of the gene considered come additionally into play at this level.[2] Finally, the presence and activity of a gene product may be directly associated with a *phenotypic function* at the organism or population level. Typically, only some aspects of molecular or cellular function are in the reach of sequence analytic studies.

Concept No. 1: Function Inheritance from a Common Ancestor Gene

The most widely known, the evolutionary (historic) approach for inferring protein function with nonexperimental means is based on the frequent observation of similarity between biomolecular sequences coding proteins with similar molecular function. Since the early examples were typically metabolic enzymes or transporters (such as hemoglobin) for which the 3D structure was available, the insight materialized soon in the paradigm of both equal/similar three-dimensional structural fold and molecular function as a consequence of similarity of protein sequence. Within this concept, a family of homologous gene/protein sequences is hypothesized to appear evolutionary during radiation of species (rarely via horizontal gene transfer) from an ancestor gene in the founding species via multiple mutations and, sometimes, gene duplications. In this context, the closest homologue of a gene in another organism ("the same gene") with most likely the same function is called orthologue, more distantly related homologues that, probably, arise from gene duplications and might assume new functions are named paralogues. Nevertheless, distant sequence similarity as a result of functional pressure or physicochemical constraints (analogous sequences in a scenario of convergent evolution) cannot always be excluded but, from the viewpoint of protein function prediction, the evolutionary pathway is not the major issue.

Functional annotation available from experimental studies of one family member is thought to be fully or partially transferable to all other members in the family. Therefore, considerable research effort has been focused on method development for more and more distantly related homologue detection to increase the likelihood of having experimentally studied family members. Except for obvious alignments with high sequence identity, it is not trivial to decide whether the similarity between sequences is significant in a statistical sense. The sequence homology approach is unthinkable without a mathematical function for measuring the similarity of two sequences quantitatively; i.e., a distance metric for the sequence space.

At the level of nucleic acids (genes and transcripts), the only possible measure is the count of identical positions in an optimal alignment. In this way, only relatively close sequence neighbors can be detected. Whereas the transcript sequence itself is just a redundant four-letter text, the translation into an amino acid sequence yields a more informative 20-letter message that often can be directly interpreted in physical and structural terms. Matrices of likelihood of amino acid type exchanges have been determined from experimentally established sequence families of globular proteins including some representatives with known tertiary structure. For example, amino acid type exchanges without changes of residue polarity/hydrophobicity or secondary structural preference impair protein structures less and are, therefore, more likely. Typically, such an exchange matrix enters the pairwise sequence similarity score function together with an empirical expression for the evaluation of evolutionary costs of deletions/insertions. For convenience of statistical evaluation, the score is recalculated into the probability (E-value) of incidentally reaching an alignment with the same or better score with a sequence taken randomly from a database of the same size. If this E-value is low, the predicted alignment is considered statistically significant. As probabilities, E-values should be always smaller than or equal to unity but analytically simplified computations of E-values, for example in the BLAST suite,[6] may lead to meaningless results above one for nonsignificant alignments with a low similarity score.

When a group of related proteins is known, then profiles that describe the likelihood of amino acid type occurrence at alignment positions can be extracted (see Step 5 in the Recipe below for detail). In turn, they allow the determination of ever more distantly related homologues in iterative cycles of profile extraction from growing alignments. Modern sequence profile techniques are the 'super-weapon' for collecting families of distantly related homologues and for assigning functions to globular domains via annotation transfer. Application of this technique lead to a number of breakthroughs in biology essentially with theoretical data analysis alone; (e.g., see refs. 7-14).

Limitations of the Homology Search Concept

The deduction of the sequence distance metric has consequences for the applicability of homology searches in databases, for example with the BLAST/PSI-BLAST suite:[6,15]

1. The sequence distance metrics have been derived from alignments of globular proteins; more accurately, from alignments of secondary structural elements (e.g., BLOSUM62[16,17]). Obviously, such similarity functions may fail for other types of sequences; for example, for

cases having *amino acid compositions* that differ drastically from those of globular proteins. For example, long hydrophobic stretches with many transmembrane regions regardless of origin have a general tendency to appear similar. The same problem create long polar runs, sequences with systematic periodicities (coiled coils, collagen, etc.) as well as sequence segments with many cysteines, prolines or tryptophanes, amino acid types that are typically rare in globular proteins and the match of which is given high weight in the similarity measure. Thus, a sequence needs to be preprocessed to filter out all probable nonglobular segments before its submission to homology searches in sequence databases. Essentially, the term "distantly related sequence homologue" is not really applicable for nonglobular regions.

2. Each alignment position contributes a summand in the total score independent of all other position. Thus, the *mutual independence of sequence positions* in their mutation ability is assumed in contrast to well-known examples of correlated mutations not only in globular proteins[18-20] but also in some shorter motifs.[21,22] Thus, sequences that fit alignments somehow at all positions but do not comply with yet hidden inter-positional constraints may nevertheless pass the sequence similarity significance criterion. This effect is practically not important for long regions of homology since the number of correlating sites is small compared with the length. In contrast, this is one of the reasons why hits with shorter alignment length are often false.
3. Yet another problem is created by the modular structure of proteins that results from *sequence segment recombination* at the genomic level. Often, the homology relationship exists rather at the level sequence segments than for whole proteins. Therefore, it becomes important to delineate these homology segments and collect their families individually.
4. *Alignment length* and *sequence identity* are of critical importance for the transferability of functional annotation. Only about 50 positions and more allow reliably assuming similarity in 3D structure.[23] With decreasing sequence identity (especially below 40%), attributes such as enzyme class, binding sites or cellular function can be transferred only with caution.[24]

Concept No. 2: Lexical Analysis, Physical Interpretation and Sequence Motif-Function Correlations

A biomolecular sequence may be analyzed in the same way as a text in a foreign language by studying occurrences/absences of certain letters (amino acid types) in the total sequence and in subsegments, by analyzing combinations of letters as well as their relative order, especially the repetitions of clusters of letters. As simple as the arithmetics of pure letter occurrences may appear, important conclusions can be drawn from such a study. The results receive a biological interpretation with the knowledge of physicochemical properties of amino acids and oligopeptides. For example, long stretches of hydrophobic amino acids may indicate secondary structural elements buried intramolecularly, within protein complexes or in lipid membranes. Runs with many polar residues are likely not to have the potential to form a hydrophobic core for a tertiary, native structure. The general relationship of hydrophobic and hydrophilic residues in larger segments might be, at least qualitatively, informative with respect to solubility and total charge. Such information can be helpful for the design of deletion mutants since those consisting mostly of hydrophobic segments are likely to produce false positive hits in a yeast two-hybrid screen and to aggregate after over-expression.

The concept of compositional bias towards certain amino acid types can be generalized with the notion of sequence complexity (information content, sequence entropy) as implemented, for example, in the SEG program.[25] Low complexity regions (LCRs) are common in sequence database proteins (~25% of all residues in sequence databases).[26,27] Sometimes, LCRs compose almost the whole protein as in the case of brakeless, a protein important for optical axon guidance in *D. melanogaster*.[28] Despite their wide spread and expected functional importance, the characterization of many LCRs, especially of those with many polar residues, still remains poor.[26]

LCRs are almost absent in known 3D structures of globular proteins (~0.5% of all residues in the protein structure database).[26,27] Thus, the concept of sequence complexity is a powerful quantitative measure for the distinction between globular (typically high complexity) and nonglobular (low complexity) regions (see ref. 29 for review). Only the high complexity regions represent good targets for sequence homology searches in database.

Many biological properties (helical transmembrane regions, coiled coils, N-terminal targeting signals, several posttranslational modifications, etc.; see Step 3 in the Recipe) are predicted from sequence with knowledge-based predictors: From a learning set of protein sequences, which are known to possess a biological feature, the encoding sequence pattern is extracted in a mathematically formalized way. Then, this pattern is searched for in query sequences, a concordance score is calculated and, in the most advanced techniques, the probability of false positive prediction is calculated. The quality of the predictor depends, first of all, on the learning set. Sometimes, it is small and does not reflect the true sequence variability in the pattern. Also, the various proteins in the learning set are typically not of the same quality with respect to their experimental verification status.

When the number of known sequences was small, a number of properties encoded in protein sequences could be associated with short amino acid type motifs ('sequence words'), which have been collected in databases, for example in PROSITE.[30] Today's sequence databases populate the available sequence space much more evenly. Therefore, short sequence motifs have a dramatically reduced predictive power (for example, the N-terminal myristoylation,[31] see also Step 2 in the Recipe).

A Recipe for Analyzing Protein Sequences

The following section is a description of a series of steps that, if executed sequentially, will typically lead to insight into structural and functional features associated with an otherwise uncharacterized protein sequence if this is achievable with existing techniques at all. With our comments below, we want to show what is generally possible but also where are the today's limits and where we have to settle for lesser goals until methodical advances move the horizons further. As practical illustration of the recipe, we invite the reader to repeat the analysis of the pds5p sequence[32] together with us (see also Fig. 1). To avoid spoiling of the text with many WWW links that change anyhow with time, this information has been collected in a regularly maintained WWW-page associated with this article (http://mendel.imp.univie.ac.at/RECIPE/).

The basic paradigm in protein sequence analysis requires the dissection of the total sequence into segments (regions, domains), each of which has its own molecular functional features. The function of the whole protein is then obtained as superposition of the segments' elementary functions.

Functional sequence regions of a protein can be classified with respect to their intrinsic structural preference in a physiological environment. Some segments have a native structure (globular domains, nonglobular helical regions in coiled coil and transmembrane regions, collagens etc.); others have not. This distinction is critical for assessing interaction capabilities: Segments with intrinsic structural preference can supply specific, stable surface recognition sites for interactions with ligands (therefore, they have a large variety of specific functions); unstructured regions cannot. As we have seen above, various types of segments require different methods for their analysis. First, nonglobular regions (phase one, steps 1-3) and, then, segments belonging to already known families of globular domains (phase two, step 4) are determined. Finally, the remaining segments are expected to represent yet unknown globular domains and are subjected to sequence family search procedures (phase three, step 5). The final step involves analysis and synthesis of the sequence analytic findings.

Step 1: Linguistic Analysis

At the beginning, it is necessary to check for linguistic particularities in the query protein sequence or its fragments. Such a textual analysis can be carried out by visual inspection or with

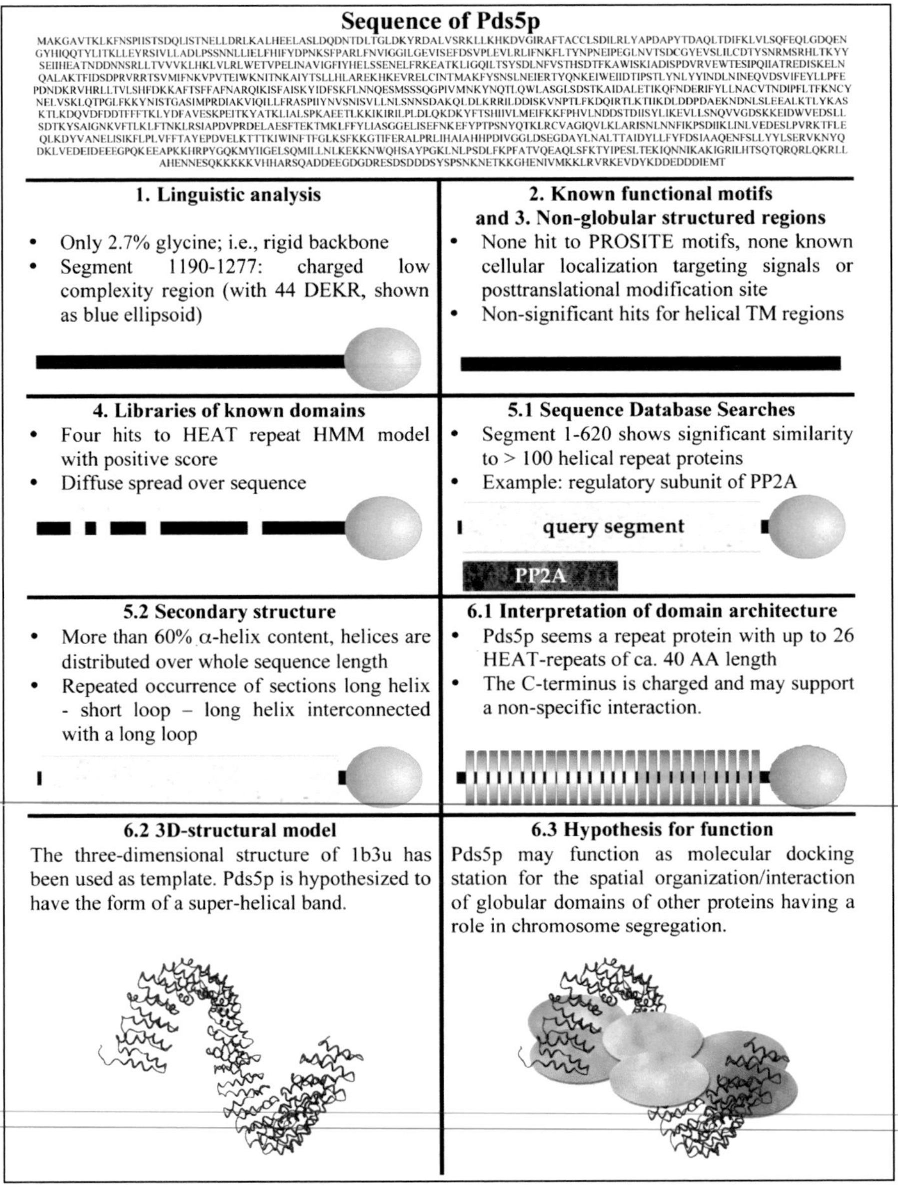

Sequence of Pds5p

MAKGAVTKLKFNSPIISTSDQLISTNELLDRLKALHEELASLDQDNTDLTGLDKYRDALVSRKLLKHKDVGIRAFTACCLSDILRLYAPDAPYTDAQLTDIFKLVLSQFEQLGDQEN
GYHIQQTYLITKLLEYRSIVLLADLPSSNNLLIELFHIFYDPNKSFPARLFNVIGGILGEVISEFDSVPLEVLRLIFNKFLTYNPNEIPEGLNVTSDCGYEVSLILCDTYSNRMSRHLTKYY
SEIIHEATNDDNNSRLLTVVVKLHKLVLRLWETVPELINAVIGFIYHELSSENELFRKEATKLIGQILTSYSDLNFVSTHSDTFKAWISKIADISPDVRVEWTESIPQIIATREDISKELN
QALAKTFIDSDPRVRRTSVMIFNKVPVTEIWKNITNKAIYTSLLHLAREKHKEVRELCINTMAKFYSNSLNEIERTYQNKEIWEIIDTIPSTLYNLYYINDLNINEQVDSVIFEYLLPFE
PDNDKRVHRLLTVLSHFDKKAFTSFFAFNARQIKISFAISKYIDFSKFLNNQESMSSSQGPIVMNKYNQTLQWLASGLSDSTKAIDALETIKQFNDERIFYLLNACVTNDIPFLTFKNCY
NELVSKLQTPGLFKKYNISTGASIMPRDIAKVIQILLFRASPIIYNVSNISVLLNLSNNSDAKQLDLKRRILDDISKVNPTLFKDQIRTLKTIIKDLDDPDAEKNDNLSLEEALKTLYKAS
KTLKDQVDFDDTFFFTKLYDFAVESKPEITKYATKLIALSPKAEETLKKIKIRILPLDLQKDKYFTSHIIVLMEIFKKFPHVLNDDSTDIISYLIKEVLLSNQVVGDSKKEIDWVEDSLL
SDTKYSAIGNKVFTLKLFTNKLRSIAPDVPRDELAESFTEKTMKLFFYLIASGGELISEFNKEFYPTPSNYQTKLRCVAGIQVLKLARISNLNNFIKPSDIIKLINLVEDESLPVRKTFLE
QLKDYVANELISIKFLPLVFFTAYEPDVELKTTTKIWINFTFGLKSFKKGTIFERALPRLIHAIAHHPDIVGGLDSEGDAYLNALTTAIDYLLFYFDSIAAQENFSLLYYLSERVKNYQ
DKLVEDEIDEEEGPQKEEAPKKHRPYGQKMYIIGELSQMILLNLKEKKNWQHSAYPGKLNLPSDLFKPFATVQEAQLSFKTYIPESLTEKIQNNIKAKIGRILHTSQTQRQRLQKRLL
AHENNESQKKKKKVHHARSQADDEEGDGDRESDSDDDSYSPSNKNETKKGHENIVMKKLRVRKEVDYKDDEDDDIEMT

1. Linguistic analysis

- Only 2.7% glycine; i.e., rigid backbone
- Segment 1190-1277: charged low complexity region (with 44 DEKR, shown as blue ellipsoid)

2. Known functional motifs and 3. Non-globular structured regions

- None hit to PROSITE motifs, none known cellular localization targeting signals or posttranslational modification site
- Non-significant hits for helical TM regions

4. Libraries of known domains

- Four hits to HEAT repeat HMM model with positive score
- Diffuse spread over sequence

5.1 Sequence Database Searches

- Segment 1-620 shows significant similarity to > 100 helical repeat proteins
- Example: regulatory subunit of PP2A

query segment

PP2A

5.2 Secondary structure

- More than 60% α-helix content, helices are distributed over whole sequence length
- Repeated occurrence of sections long helix - short loop – long helix interconnected with a long loop

6.1 Interpretation of domain architecture

- Pds5p seems a repeat protein with up to 26 HEAT-repeats of ca. 40 AA length
- The C-terminus is charged and may support a non-specific interaction.

6.2 3D-structural model

The three-dimensional structure of 1b3u has been used as template. Pds5p is hypothesized to have the form of a super-helical band.

6.3 Hypothesis for function

Pds5p may function as molecular docking station for the spatial organization/interaction of globular domains of other proteins having a role in chromosome segregation.

Figure 1. Sequence analysis of yeast pds5p. When the sequence-analytic study of yeast pds5p was started, only its sequence (top of the figure, 1277 amino acid residues) and its knock-out phenotype in mitosis were known.[32] Searches for nonglobular regions detected only a strongly charged region at the C-terminus. Compositional studies revealed a surprisingly low content of glycines indicating a generally rigid backbone. Three arguments (comparison with known domain profiles of helical repeats, distant similarity with the regulatory subunit of PP2A and predicted helical secondary structure including also the pattern of two helices interconnected by a short loop and a long loop between helix pairs) support the view that HEAT-repeats occupy the major part of the sequence. The reliability of these predictions decreases towards the C-terminal part. The HEAT-repeat region is suggested to fold into a super-helical band with interaction sites for other proteins, the charged C-terminal region has, apparently, a role for unspecific amplification of some binding reaction.[32]

computerized tools such as SAPS.[33] This program incorporates also rigorous statistical criteria for finding significant differences of the query's lexical properties from averages of SWISS-Prot sequences.[34]

Regions of low sequence complexity, another important lexical property, can be determined with tools such as SEG[25] or CAST.[35] The SEG program has three recommended parametrizations with sequence windows of w = 12, 25 or 45 residues. In standard applications, only the smallest window, the most stringent criterion, is applied. Personal experience shows that the larger window (w = 25) helps detecting less obvious LCRs, although SEG marks sometimes also globular regions as LCRs if applied with maximal window size (w = 45). The final output of SEG should be preprocessed for further analysis: (1) Sometimes, SEG leaves a small segment (with length below window size) between two neighboring LCRs unassigned. Such a segment can often be fused with the two LCRs into a single larger LCR. (2) Evaluation of polarity of LCRs is helpful for their functional assessment. Hydrophobic LCRs (rarely longer than 30 residues) often have a role in membrane attachment or are buried internally in protein complexes. Functional assignment of polar LCRs, especially those with more than 100 residues length, is more problematic. Polar LCRs are thought to be intrinsically unstructured and in contact with the aqueous phase. Some serve as mechanical linkers between domains, have a role in electrostatic interactions or carry sites for posttranslational modifications. The specific molecular function of polar LCRs is typically unclear except for rare cases.[26,28,29]

Step 2: Motifs for Subcellular Targeting and Posttranslational Modifications

A number of functional motifs for posttranslational modifications or targeting to subcellular localizations are located within sequence regions without intrinsic structural preference. Specialized predictors can test the occurrence of these motifs. Several N-terminal signals involving typically 20-40 residues encode targeting to organelles: SIGNALP[36] recognizes the signal leader peptide for export to the endoplasmic reticulum. CHLOROP[37] searches for chloroplast- and another tool[38] for mitochondrion-targeted proteins. SIGNALP in its recent version has very reasonable prediction accuracy above 80% for true predictions for large sequence sets and a low rate (~14-19%) for false-positive hits and compares favorably with alternative tools.[39] Prediction of chloroplast- and mitochondrial targeting are not comparable in this respect, first of all, because the available sets of experimentally learning sequences are less comprehensive and reliable. TARGETP[40] represents a unified version of all three predictors. A new predictor for the C-terminal PTS1 signal (with a length of about 12 residues) that encodes pex5-dependent peroxysomal localization has a sensitivity >95% and a selectivity below 0.5%.[41]

Several lipid posttranslational modifications of proteins can now be reliably predicted from sequence. (1) N-terminal N-myristoylation is encoded by a signal of about 17 residues. It is recognized with >95% for true sites and with less than 0.5% for unrelated sequences by a recently developed tool.[22,31] In some cases of posttranslational processing, internal glycines become N-terminal and myristoylated. This program analyzes also a number of such scission patterns. N-terminal N-myristylation with subsequent palmytoylation (if there are cysteines close to the N-terminus) might hint at a noncanonical export mechanism.[42] (2) Glycosylphosphatidylinositol (GPI) lipid anchoring is a posttranslational modification of protein C-termini carrying the respective recognition signal of ca. 40 residues. The anchor is attached after proteolytic scission of a propeptide. The big-II predictor predicts GPI lipid anchor attachment (~80% accuracy for truly anchored animal proteins with ~0.2% false positives) and computes also the one or two most probable attachment sites.[21,43-47] (3) A recently released predictor for farnesylation and geranylgeranylation, the two types of prenylation at protein C-termini, is accessible from the WWW-page associated with this article.

The localization and lipid modification signals discussed above involve 12-50 residues from the respective termini. Typically, they are not characterized by amino acid type preferences alone but also by sequence context involving a strong pattern of physical properties and, partially, by some inter-positional correlations within the motif. Only this additional information allows reliable motif detection in uncharacterized sequences and the assessment of the possible prediction error in tests involving dozens of thousands of sequences.[48]

It should be emphasized that conservation of a handful of residues in a short motif alone does not imply function correlation and, barely, supports more than a working hypothesis. Typically, short, polar oligopeptides do not have intrinsic structural preferences;[49] they cannot supply a stable interface for intermolecular interactions. Even in the case of true function embedding into an unstructured region of a protein that interacts with a globular domain of another protein, a functional motif requires a sequential environment involving residues for less specific interactions and linker function.[21,22,48]

To illustrate, a number of short PROSITE motifs[30] are also used for characterizing post-translational modification sites (for example, for phosphorylation, N-glycosylation and myristoylation) but with a high rate of false hits.[31] Other arguments (e.g., experimental data) are needed to support the relevance of predicted sites. There are alternative neural network based predictors for phosphorylation,[50] O- and N-glycosylation[51,52] but their prediction accuracy is not yet sufficient for unsupervised sequence annotation.

Similarly, many other, scarcely described and yet insufficiently understood sequence signals, e.g., for nuclear import[53] and export[54] or the PEST degradation signature,[55] circulate widely in the literature but their predictive significance for sequence analysis is still low since the correlation between protein sequence variability and function remains ambiguous. Often, the biological mechanisms for read-out of these signals are poorly understood.

Step 3: Nonglobular Regions with Intrinsic Structural Preference

At early stages of sequence studies, it is important to recognize α-helical transmembrane regions and coiled coil segments. Both have compositional bias, which is often not recognized by sequence complexity computing programs, and, consequently, these segments should also be removed from the sequence before submission to searches for distant relatives in sequence databases.

Coiled coil regions can be predicted from sequence with the updated COILS algorithm of Lupas.[56] Typically, WWW-server versions run COILS only with standard parametrization and, sometimes, predict coiled coils wrongly in regions with many polar residues without any hydrophobic amino acids in 'a' and 'd' positions of the heptade repeat. A second COILS run with a changed weighting for polar residues as recommended in the manual diagnoses many of those doubtful assignments. To notify, there are also versions of COILS in the public domain erroneously deviating from the original implementation of algorithm and resulting in fewer and shorter predicted coiled coil segments for some proteins.

There may be other fibrillar segments in proteins. For example, collagen segments are recognized by typical glycine- and proline-rich repeats and this property is incorporated in an HMM of the PFAM domain PF01391.[57]

The prediction of membrane attachment of integral membrane proteins via protein segments immersed into the lipid bilayer is still problematic. If transmembrane helical regions are present, they are readily recognized by prediction tools like TMHMM[58] or DAS-TMfilter, a recent update of DAS,[59] as well by a number of other programs.[60] With less accuracy, the protein topology with respect to the membrane is predicted (mostly based on the positive-inside-rule[61]). Since the motif description rests almost entirely on the requirement of long hydrophobic stretches (except for a minimum length), false positive prediction, especially of single membrane-pass proteins is frequent. TMHMM and DAS-TMfilter have a better selectivity than the competing programs but they also fail for proteins with long helical, hydrophobic repeats (for example, ARM/HEAT repeat proteins such as tis7 (gi321269) or inscuteable (gi1079094)).

The architectural diversity of proteins attached to membranes involves more than just transmembrane helical regions but these configurations cannot be predicted with available TM region prediction tools. For example, there is an interesting class of amphipatic helices embedded into the membrane parallel to the bilayer surface (monotopic membrane proteins).[62-64] Further, transmembrane helix formation is not entirely determined locally by the hydrophobic stretch itself but may depend on the rest of the protein sequence[65] or even complex formation.

Step 4: Known Sequence Families of Globular Domains

Globular domains are the main structural and functional building blocks of proteins. Various definitions of the notion 'domain' differ but their content is overlapping. From the viewpoint of three-dimensional structure, a domain is a compact, spatially distinct unit with its own hydrophobic core, the fundament of its native tertiary structure. In the kinetic sense, a native structure implies that conformational fluctuations are locally confined (i.e., are smaller than the size of the three-dimensional structure). Thus, globular domains can supply stable interfaces and recognition sites for other molecules, even for those without intrinsic structural preference. Thermodynamically, a domain is melting independently. Often, a domain is considered an autonomous folding unit. At the same time, a structural unit might not be continuous in the sequence. In the evolutionary perspective and in sequence comparisons, a domain is a family of significantly similar sequences that are related by their mutational history. From the functional viewpoint, domains may be promiscuous with different active sites and binding capabilities for various sequence family members but the degree of diversity is uneven among domains. A typical globular domain involves 100-150 amino acid residues;[66-68] thus, much longer segments can be supposed to involve several independent domains. To avoid confusions, it is advised to use the term "domain" in the sense of globular domain and to apply sequence region or segment in other context.

At this stage of analysis, it is a good decision to compare the target sequence with entries in public domain databases. There are traditional profile-based (PROSITE,[30] BLOCKS,[69] PRINTS[70]); hidden Markov model (HMM)-based (PFAM,[57] SMART,[71] significance threshold typically E~0.1); combined tools (PANAL[72]) and RPS-BLAST profile-based (CDD search,[73] significance threshold typically E~0.01) collections. There are at least two reasons: The given sequence might be so distantly related to a known family that a simple pairwise similarity search with the query or any of the family members would not detect that relationship. Profiles describing whole families are much more sensitive. Second, one domain in multi-domain targets may have so many close relatives in the sequence database that the output list from a BLAST search with the full sequence would be obliterated with those hits alone. It makes sense to compare a query with all available domain libraries since definitions of even actually the same domain may slightly differ and numerical noise can lead to hits in one but not in another library.

Currently, there are two major primary domain libraries. PFAM is unprecedented in sequence coverage.[57] At the same time, the domain definitions may contain slight inconsistencies mostly concerning boundaries of domains. Sometimes, signal peptides, fibrillar protein segments or helical transmembrane regions are included into the profile or the domain definition contains actually several domains. SMART is a very carefully curated but much smaller domain databases that focuses on certain classes of signaling, nuclear and extracellular proteins.[71] SMART domain boundaries typically define the core of a single globular domain.[74]

There are two modes for searching the occurrence of domains in query sequences with HMMs and profiles. In the so-called global mode, the presence of only complete domains is assumed and the optimal alignment of a query segment with the complete domain profile is searched. This mode is typically more sensitive that the fragmented domain search where also partial hits of the domain profile in the query are reported. In the ideal case, both regimes deliver the same result. Most hits from the fragmented domain search are meaningless in the absence of full-domain matches but if they coincide with known binding sites for ligands or otherwise functionally relevant parts of the domain, careful sequence inspection may lead to a discovery of very distantly related sequence homologues.

A fragmented domain search with the profile of the histone acetyltransferase family has hit eco1p, a yeast protein for the establishment of cohesion between chromatids during mitosis, in the region of the acetyl-CoA binding site.[9] The partial hit was extended with arguments based on secondary structure prediction and the conservation of a hydrophobic pattern. This finding stimulated experimental analysis and finally led to the discovery of a new family of acetyl-CoA binding and acetyl-transferring enzymes with a role in cohesion.[9]

The domains with multiple internal structural repeats are difficult to detect; therefore, this domain class requires special attention.[75] Such repeats are known as closed structures (e.g., β-propellers) or as semi-closed forms, for example the superhelical armadillo or heat repeats. Many repeat proteins have scaffolding functions for protein-protein interactions. For repeat detection, the query should be cleaned from compositionally biased regions in accordance with steps 1-3 of the recipe. The PROSPERO tool[76] is designed for recognizing even subtle internal sequence repeats. Since it operates with rigorous statistical criteria, the validity of the finding can be assessed in probabilistic terms. The REP tool[77] compares the query sequence with an HMM library of known repeats. Unfortunately, the evolutionary pressure for sequence conservation within repeats is typically low and reduced to the requirement of packing and maintenance of the hydrophobic core. Therefore, even hits with low statistical significance deserve attention.

Step 5: Sequence Database Searches

Searches for similar sequences in databases can be applied in two different contexts. Full sequence searches are reasonably aimed only at finding closely related sequential neighbors where the methodical details of deriving the sequence distance metric do not have a major impact on the search result (typically, log (*E - value*) < - 10 for BLAST).

A search in sequence databases for similar but distantly related proteins with the target under study is in fact the last step of sequence analysis. Only sequence segments without low complexity, transmembrane and coiled coil regions, peptide segments for posttranslational modifications and cellular targeting, and known domains can routinely be subjected to such searches. Now, the effort is aimed at collecting the complete sequence family. The larger the family, the higher is the probability of hitting a functionally annotated family member. Additionally, it is necessary to understand the sequence variability within the sequence.

Traditionally, this a process of repeated application of pairwise sequence comparison techniques such as BLAST and general profile-searching techniques relying on manually or automatically constructed alignments (PSI-BLAST[78] with inclusion E-values up to ~0.01, SAM-T99,[79,80] or a combination of Clustalx[81,82] with a profile searching technique). Both the primary query as well as any new family members is subjected to such searches. The optimal search heuristics are a matter of continued scientific discussion.[74] Large sequence families have an internal structure consisting of clusters of sequentially (and, often, functionally) more similar proteins with statistically significant links between them.

Three aspects deserve additional comments: First, borderline hits require visual inspection before inclusion into the family or their final rejection. An excellent review of physical and structural criteria for nonstatistical evaluation of alignment significance (based on considerations of protein structural architecture) has been supplied by Bork and Gibson.[83] Reoccurrence of some motif conserved within the family might indicate correct assignment. Finally, the correct inclusion into the family should be verified by a reciprocal database search (started with the doubtful sequence segment) that collects already verified family members with statistical significance. It must be noted that many database search programs are not 100% commutative with respected to starting and hit sequences due to algorithmic simplifications that save computing time. Second, manually constructed alignments may be superior over those automatically generated, especially if 3D structural information for at least one family member is available. In the case of the pleckstrin homology (PH) domain sequences, sequence identities had been very low but reliable alignments applicable for further rounds of profile searches were obtained with manual adjustment emphasizing the conserved hydrophobic patterns and a conserved tryptophane position.[84,85] Third, the probability of finding hits can be increased if EST and genome databases are six-frame translated on the fly and included into the search for relatives.[86] In a few cases, some relaxation of search thresholds leads to the necessary intermediate sequence hits during family collection. Fourth, since most amino acid substitution matrices give high weights to matches of rare residues such as cysteine, database searches with such a sequence segment to database searches may result in spurious hits with underestimated E-values, which may become close to standard selection thresholds. This has happened in the case of the C-terminal domain of wingless/wnt-1 that was incorrectly suggested to be related to the lipid-binding domain of phospholipase A2.[87] This possibility was later ruled out by structural arguments (completeness of the hydrophobic core, satisfaction of disulphid bonds).[88]

Until recently, it was very difficult to find routinely so distantly related family members with known 3D structures that have no recognizable sequence similarity with pure sequence-based approaches but, nevertheless, have the same fold. Higher sensitivity is achieved in comparisons of two profiles, one extracted from the query's sequence family and the other from a family of proteins of similar 3D structure and their sequential homologues. In addition to information from amino acid letter comparison, some structural information can be mobilized: The alignment of query sequences with structural templates, the mapping of sequence positions to structural positions, allows, for example, scoring of the agreement between predicted secondary structure of the query with the secondary structure of the template or the polarity of amino acid residues of the query with the accessibility of template sites. Different strategies have been implemented in 3D-PSSM,[89] bioinbqu,[90] DOE FOLD predictor,[91] FFAS,[92] PSIPRED,[93] SAM,[79] SDSC1[94] and SUPERFAMILY,[95] which are available as WWW-servers. Generally, their predictions have to be viewed with caution. Similar predictions for various sequence family members are indicative for higher significance. Some of these techniques have been equipped with methods for assessing the probability of false positive prediction. There are cases where the prediction of the 3D-structure with fold predictors has produced the decisive hint. For example, the predicted β-propeller structure of the globular domain of PIG-T can explain its molecular function as gate mechanism for protein substrates of the transamidase PIG-K in the GPI lipid anchor biosynthesis pathway.[96]

Yet another approach for enlarging the sequence family focuses on sequence architecture, the linear order of functional segments in a protein. Sub-threshold similarity in some sequence segment combined with similar length and order of other architectural elements can indicate on the existence of homologues in other species, even if the evolutionary divergence has become high.[97,98]

After having the sequence family completed, the family sequence alignment, known structures of family members, the available sequence annotation and the scientific literature for all family members have to be studied. First, conservation patterns of hydrophilic/hydrophobic residues and of secondary structural elements (indicating fold conservation), or of motifs with functional residues (giving a hint at conserved ligand binding and active sites) have to be taken into account.[99] The secondary structure predicted with JPRED[100] or PSIPRED[93] for the sequence family can help in the interpretation of the data. Second, details of known structures of family members that do not depend on the sequence-variable positions should be searched for. For example, the distribution of the electrostatic potential at the protein surface is sometimes invariant within a family and may explain the binding behavior. Searches for proteins with the same fold[101] can give lead to functional information on other proteins with the same fold. Third, the taxonomic distribution[102] of the family is informative with respect to the evolution of the cellular processes involving the sequence domain studied. Sometimes, evolutionary trees constructed from all family members may yield additional insight.

The scientific literature must be searched for experimental evidence of biological function that can be linked with the sequence segment in some family members. The degree of possible annotation transfer from family members to the target under consideration depends on many aspects. As a rule, the similarity with respect to the 3D fold can be determined with greater reliability but molecular and, the more, cellular functional descriptors cannot always be transferred with the same confidence due to considerable plasticity of protein function.[5,103] For example, a large family of proteins has in common a domain responsible for ras-binding in the case of many family members.[104] This information was extrapolated to the whole family including the Rho-GTPase-activating protein myr-5. For the latter one, it turned out that the presence of this domain and its fold was predicted correctly, but the actual function was not.[105]

Step 6: Analysis of Sequence Analytic Findings and Synthesis of Molecular Function

First, it is necessary to evaluate the reliability of predictions and annotations for overlapping sequence segments and to resolve possible contradictions. Then, the prediction results should allow segmentation of the query sequence into sequence regions, to which the collected structural and functional annotation can be attached. Often, some experimental data for the protein analyzed is available from the cooperating experimental researchers, which has to be discussed now in context with the sequence-analytic findings. Synthesis of the segments' functions into the protein function is the most creative step in the whole procedure where the biological knowledge of the researcher and his experience in using sequence analytic methods come together. It is possible that the collected evidence is so strong that there is no doubt (see ref. 106 for discussion). In most cases, the thought concentrates on consequences for the further experimental strategy. For example, clear directives can be given for mutant design: Deletion mutants should follow the derived segmental structure; point mutation should focus on conserved sequence positions.

Protein structure and function are encrypted in the protein sequence; thus, they can be predicted relying on amino acid sequence information in principle. Sometimes, molecular and cellular properties can be predicted. Phenotypic functions are usually outside the predictive power of sequence analytic studies (only in cases of clear homology). It should be emphasized that there are aspects of molecular function that strongly resist theoretical treatment. It is highly unlikely that theoretical methods will predict biological features without any analogy to experimentally studied cases since all procedures finally rely on observed sequence-function correlation.

Even if the 3D structures of two individual subunit proteins are known, it is still not possible to reliably predict the specific protein-protein interaction in a complex.[107] In the general case, there is no way to predict even the fact of complex formation from sequence alone. Potential hints can be obtained from homology considerations but, as in the case of the putative ras-binding activity of myr-5,[104,105] with low reliability. Sometimes, conservation of gene order or regulatory genomic neighborhood, gene fusion events or the conserved cooccurrence of genes in different genomes might be supportive for interaction[108-110] but not more. With large-scale mass spectrometric analysis, list of proteins in complexes have been compiled that can be looked up as well as interactions from two-hybrid screens.[111-114]

Concluding Remarks

The development of high-throughput experimental technologies and its first major breakthrough, the complete sequencing of the genomes of organisms ranging from viruses over bacteria, lower eukaryotes to human, has changed life science research qualitatively. For the first time, the biological object can be studied in its totality at the molecular level. The immediate task for the coming decade consists in assigning functions to all genes known by sequence. Since the new data are so large and their the biological interpretation require complex approaches, theoretical science can and must contribute decisively to the research progress. The research success in life sciences depends increasingly on the ability of researchers in experimental and theoretical biology to jointly focus on relevant questions.

Modern protein sequence analysis relies on two major approaches: protein homology searches based on the concept of statistically significant sequence similarity and textual analysis with physical interpretation and the extrapolation of empirical relationships established between local sequence motifs and patterns with structural and functional properties.

Acknowledgements

The author is grateful for generous support from Boehringer Ingelheim and from the Bioinformatics Network within the Genome Research Austria program (Gen-AU BIN). F. Eisenhaber thanks Peer Bork (EMBL Heidelberg) who attracted him to the field of biomolecular sequence analysis during 1996 and readily shared his practical experience. This chapter benefited from conversations with Chris Ponting (MRC Oxford) and Eugene Koonin (NCBI Bethesda) as well as from interactions with wet lab IMP scientists that kindly shared their findings with the IMP bioinformatics group on a continuous basis. Special thanks are to B. Eisenhaber, W. Kubina, S. Maurer-Stroh, G. Neuberger, M. Novatchkova, A. Schleiffer, G. Schneider, S. Washietl, M. Wildpaner for discussions of various aspects of this work and S. Maurer-Stroh for support in creating parts of Figure 1.

References

1. Novatchkova M, Eisenhaber F. Can molecular mechanisms of biological processes be extracted from expression profiles? Case study: Endothelial contribution to tumor-induced angiogenesis. Bioessays 2001; 23:1159-1175.
2. Zhang MQ. Computational prediction of eukaryotic protein-coding genes. Nat Rev Genet 2002; 3:698-709.
3. Fickett JW. ORFs and genes: How strong a connection? J Comput Biol 1995; 2:117-123.
4. Harrison PM, Hegyi H, Balasubramanian S et al. Molecular fossils in the human genome: Identification and analysis of the pseudogenes in chromosomes 21 and 22. Genome Res 2002; 12:272-280.
5. Bork P, Dandekar T, Diaz-Lazcoz Y et al. Predicting function: From genes to genomes and back. J Mol Biol 1998; 283:707-725.
6. Altschul S, Boguski M, Gish W et al. Issues in searching molecular sequence databases. Nature Genetics 1994; 6:119-129.
7. Yuan YP, Schultz J, Mlodzik M et al. Secreted fringe-like signaling molecules may be glycosyltransferases. Cell 1997; 88:9-11.
8. Rea S, Eisenhaber F, O'Carroll D et al. Regulation of chromatin structure by site-specific histone h3 methyltransferases. Nature 2000; 406:593-599.
9. Ivanov D, Schleiffer A, Eisenhaber F et al. Eco1 is a novel acetyltransferase that can acetylate proteins involved in cohesion. Curr Biol 2002; 12:323-328.
10. Dlakic M. Chromatin silencing protein and pachytene checkpoint regulator dot1p has a methyltransferase fold. Trends Biochem Sci 2001; 26:405-407.
11. van Leeuwen F, Gafken PR, Gottschling DE. Dot1p modulates silencing in yeast by methylation of the nucleosome core. Cell 2002; 109:745-756.
12. Aravind L, Koonin EV. The DNA-repair protein AlkB, EGL-9, and leprecan define new families of 2-oxoglutarate- and iron-dependent dioxygenases. Genome Biol 2001; 2:RESEARCH0007.
13. Trewick SC, Henshaw TF, Hausinger RP et al. Oxidative demethylation by escherichia coli AlkB directly reverts DNA base damage. Nature 2002; 419:174-178.
14. Falnes PO, Johansen RF, Seeberg E. AlkB-mediated oxidative demethylation reverses DNA damage in Escherichia Coli. Nature 2002; 419:178-182.
15. Altschul SF, Madden TL, Schaffer AA et al. Gapped blast and PSI-blast: A new generation of protein database search programs. Nucleic Acids Res 1997; 25:3389-3402.
16. Henikoff S, Henikoff JG. Amino acid substitution matrices from protein blocks. Proc Nat Acad Sci USA 1992; 89:10915-10919.
17. Henikoff S, Henikoff JG. Amino acid substitution matrices. Adv Protein Chem 2000; 54:73-97.
18. Pollock DD, Taylor WR, Goldman N. Coevolving protein residues: Maximum likelihood identification and relationship to structure. J Mol Biol 1999; 287:187-198.
19. Cootes AP, Curmi PM, Cunningham R et al. The dependence of Amino acid pair correlations on structural environment. Proteins 1998; 32:175-189.
20. Chelvanayagam G, Eggenschwiler A, Knecht L et al. An analysis of simultaneous variation in protein structures. Protein Eng 1997; 10:307-316.
21. Eisenhaber B, Bork P, Eisenhaber F. Sequence properties of GPI-anchored proteins near the Ω-site: Constraints for the polypeptide binding site of the putative transamidase. Protein Eng 1998; 11:1155-1161.
22. Maurer-Stroh S, Eisenhaber B, Eisenhaber F. N-terminal N-myristoylation of proteins: Prediction of substrate proteins from Amino acid sequence. J Mol Biol 2002; 317:541-557.
23. Sander C, Schneider R. Database of homology-derived protein structures and the structural meaning of sequence alignment. Proteins 1991; 9:56-68.

24. Devos D, Valencia A. Practical limits of function prediction. Proteins 2000; 41:98-107.
25. Wootton JC, Federhen S. Analysis of compositionally biased regions in sequence databases. Methods Enzymol 1996; 266:554-571.
26. Wootton JC. Sequences with 'Unusual' Amino acid compositions. Curr Op Struct Biol 1994; 4:413-421.
27. Saqi M. An analysis of structural instances of low complexity sequence segments. Protein Eng 1995; 8:1069-1073.
28. Senti K, Keleman K, Eisenhaber F et al. Brakeless is required for lamina targeting of R1-R6 axons in the Drosophila visual system. Development 2000; 127:2291-2301.
29. Eisenhaber B, Eisenhaber F. Sequence complexity of proteins and its significance in annotation. In: Subramaniam S, ed. Bioinformatics in the Encyclopedia of Genetics, Genomics, Proteomics and Bioinformatics. New York: Wiley Interscience, 2005:4, (DOI:10.1002/047001153X.g403313).
30. Falquet L, Pagni M, Bucher P et al. The PROSITE database, its status in 2002. Nucleic Acids Res 2002; 30:235-238.
31. Maurer-Stroh S, Eisenhaber B, Eisenhaber F. N-terminal N-myristoylation of proteins: Refinement of the sequence motif and its taxon-specific differences. J Mol Biol 2002; 317:523-540.
32. Panizza S, Tanaka T, Hochwagen A et al. Pds5 cooperates with cohesin in maintaining sister chromatid cohesion. Curr Biol 2000; 10:1557-1564.
33. Brendel V, Bucher P, Nourbakhsh IR et al. Methods and algorithms for statistical analysis of protein sequences. Proc Natl Acad Sci USA 1992; 89:2002-2006.
34. Karlin S, Brendel V. Chance and statistical significance in protein and DNA sequence analysis. Science 1992; 257:39-49.
35. Promponas VJ, Enright AJ, Tsoka S et al. CAST: An iterative algorithm for the complexity analysis of sequence tracts. Complexity analysis of sequence tracts. Bioinformatics 2000; 16:915-922.
36. Nielsen H, Brunak S, von Heijne G. Machine learning approaches for the prediction of signal peptides and other protein sorting signals. Protein Eng 1999; 12:3-9.
37. Emanuelsson O, Nielsen H, von Heijne G. ChloroP, a neural network-based method for predicting chloroplast transit peptides and their cleavage sites. Protein Sci 1999; 8:978-984.
38. Emanuelsson O, von Heijne G, Schneider G. Analysis and prediction of mitochondrial targeting peptides. Methods Cell Biol 2001; 65:175-187.
39. Menne KM, Hermjakob H, Apweiler R. A comparison of signal sequence prediction methods using a test set of signal peptides. Bioinformatics 2000; 16:741-742.
40. Emanuelsson O, von Heijne G. Prediction of organellar targeting signals. Biochim Biophys Acta 2001; 1541:114-119.
41. Neuberger G, Maurer-Stroh S, Eisenhaber B et al. Prediction of PTS1 signal dependent peroxysomal targeting from protein sequences. submitted 2002.
42. Denny PW, Gokool S, Russell DG et al. Acylation-dependent protein export in leishmania. J Biol Chem 2000; 275:11017-11025.
43. Eisenhaber B, Bork P, Eisenhaber F. Prediction of potential GPI-modification sites in proprotein sequences. J Mol Biol 1999; 292:741-758.
44. Eisenhaber B, Bork P, Yuan Y et al. Automated annotation of GPI anchor sites: Case study C. Elegans. Trends Biochem Sci 2000; 25:340-341.
45. Eisenhaber B, Bork P, Eisenhaber F. Post-translational GPI lipid anchor modification of proteins in kingdoms of life: Analysis of protein sequence data from complete genomes. Protein Eng 2001; 14:17-25.
46. Eisenhaber B, Schneider G, Wildpaner M et al. A sensitive predictor for potential GPI lipid modification sites in fungal protein sequences and its application to genome-wide studies for aspergillus nidulans, candida albicans, neurospora crassa, Saccharomyces Cerevisiae and schizosaccharomyces pombe. J Mol Biol 2004; 337:243-253.
47. Eisenhaber B, Wildpaner M, Schultz CJ et al. Glycosylphosphatidylinositol lipid anchoring of plant proteins. Sensitive prediction from sequence- and genome-wide studies for arabidopsis and rice. Plant Physiol 2003; 133:1691-1701.
48. Eisenhaber B, Eisenhaber F, Maurer-Stroh S et al. Prediction of sequence signals for lipid post-translational modifications: Insights from case studies. Proteomics 2004; 4:1614-1625.
49. Minor Jr DL, Kim PS. Context-dependent secondary structure formation of a designed protein sequence. Nature 1996; 380:730-734.
50. Blom N, Gammeltoft S, Brunak S. Sequence and structurebased prediction of eukaryotic protein phosphorylation sites. J Mol Biol 1999; 294:1351-1362.
51. Hansen JE, Lund O, Tolstrup N et al. NetOglyc: Prediction of mucin type O-glycosylation sites based on sequence context and surface accessibility. Glycoconj J 1998; 15:115-130.
52. Gupta R, Brunak S. Prediction of glycosylation across the human proteome and the correlation to protein function. Pac Symp Biocomput 2002; 310-322.

53. Cokol M, Nair R, Rost B. Finding nuclear localization signals. EMBO Rep 2000; 1:411-415.
54. Yoneda Y. Nucleocytoplasmic protein traffic and its significance to cell function. Genes Cells 2000; 5:777-787.
55. Rechsteiner M, Rogers SW. PEST sequences and regulation by proteolysis. Trends Biochem Sci 1996; 21:267-271.
56. Lupas A. Predicting coiled-coil regions in proteins. Curr Opin Struct Biol 1997; 7:388-393.
57. Bateman A, Birney E, Cerruti L et al. The Pfam protein families database. Nucleic Acids Res 2002; 30:276-280.
58. Krogh A, Larsson B, von Heijne G et al. Predicting transmembrane protein topology with a hidden markov model: Application to complete genomes. J Mol Biol 2001; 305:567-580.
59. Cserzo M, Wallin E, Simon I et al. Prediction of transmembrane alpha-helices in prokaryotic membrane proteins: The dense alignment surface method. Protein Eng 1997; 10:673-676.
60. Moller S, Croning MD, Apweiler R. Evaluation of methods for the prediction of membrane spanning regions. Bioinformatics 2001; 17:646-653.
61. von Heijne G. Membrane protein structure prediction. Hydrophobicity analysis and the positive-inside rule. J Mol Biol 1992; 225:487-494.
62. Picot D, Garavito RM. Prostaglandin H synthase: Implications for membrane structure. FEBS Lett 1994; 346:21-25.
63. Wendt KU, Lenhart A, Schulz GE. The structure of the membrane protein squalene-hopene cyclase at 2.0 a resolution. J Mol Biol 1999; 286:175-187.
64. Sukumar N, Xu Y, Gatti DL et al. Structure of an active soluble mutant of the membrane-associated (S)- mandelate dehydrogenase. Biochem 2001; 40:9870-9878.
65. Goder V, Spiess M. Topogenesis of membrane proteins: Determinants and dynamics. FEBS Lett 2001; 504:87-93.
66. Trifonov EN. Segmented structure of protein sequences and early evolution of genome by combinatorial fusion of DNA elements. J Mol Evol 1995; 40:337-342.
67. Wheelan SJ, Marchler-Bauer A, Bryant SH. Domain size distributions can predict domain boundaries. Bioinformatics 2000; 16:613-618.
68. Xu D, Nussinov R. Favorable domain size in proteins. Fold Des 1998; 3:11-17.
69. Henikoff JG, Pietrokovski S, McCallum CM et al. Blocks-based methods for detecting protein homology. Electrophoresis 2000; 21:1700-1706.
70. Attwood TK, Beck ME, Flower DR et al. The PRINTS protein fingerprint database in its fifth year. Nucleic Acids Res 1998; 26:304-308.
71. Letunic I, Goodstadt L, Dickens NJ et al. Recent improvements to the SMART domain-based sequence annotation resource. Nucleic Acids Res 2002; 30:242-244.
72. Silverstein KA, Kilian A, Freeman JL et al. PANAL: An integrated resource for protein sequence ANALysis. Bioinformatics 2000; 16:1157-1158.
73. Marchler-Bauer A, Panchenko AR, Shoemaker BA et al. CDD: A database of conserved domain alignments with links to domain three-dimensional structure. Nucleic Acids Res 2002; 30:281-283.
74. Ponting CP, Schultz J, Copley RR et al. Evolution of domain families. Adv Protein Chem 2000; 54:185-244.
75. Chelvanayagam G, Knecht L, Jenny T et al. A combinatorial distance-constraint approach to predicting protein tertiary models from known secondary structure. Fold Des 1998; 3:149-160.
76. Mott R. Accurate formula for P-values of gapped local sequence and profile alignments. J Mol Biol 2000; 300:649-659.
77. Andrade MA, Ponting CP, Gibson TJ et al. Homology-based method for identification of protein repeats using statistical significance estimates. J Mol Biol 2000; 298:521-537.
78. Altschul SF, Koonin EV. Iterated profile searches with PSI-BLAST—a tool for discovery in protein databases. Trends Biochem Sci 1998; 23:444-447.
79. Karplus K, Hu B. Evaluation of protein multiple alignments by SAM-T99 using the BAliBASE multiple alignment test set. Bioinformatics 2001; 17:713-720.
80. Karplus K, Karchin R, Barrett C et al. What is the value added by human intervention in protein structure prediction? Proteins 2001; 45(Suppl 5):86-91.
81. Thompson JD, Higgins DG, Gibson TJ. CLUSTAL W: Improving the sensitivity of progressive multiple sequence alignment through sequence weighting, position-specific gap penalties and weight matrix choice. Nucleic Acids Res 1994; 22:4673-4680.
82. Higgins D, Thompson JD, Gibson TJ. Using CLUSTAL for multiple sequence alignment. Meth Enzymol 1996; 266:383-402.
83. Bork P, Gibson TJ. Applying motif and profile searches. Meth Enzymol 1996; 266:162-184.
84. Musacchio A, Gibson TJ, Rice P et al. The PH-domain: A common piece in the structural patchwork of signalling proteins. Trends Biochem Sci 1993; 18:343-348.

85. Gibson TJ, Hyvönen M, Musacchio A et al. PH domain: The first anniversary. Trends Biochem Sci 1994; 19:349-353.
86. Aravind L, Koonin EV. Classification of the caspase-hemoglobinase fold: Detection of new families and implications for the origin of the eukaryotic separins. Proteins 2002; 46:355-367.
87. Reichsman F, Moore HM, Cumberledge S. Sequence homology between wingless/Wnt-1 and a lipid-binding domain in secreted phospholipase A2. Curr Biol 1999; 9:R353-R355.
88. Barnes MR, Russell RB, Copley RR et al. A lipid-binding domain in Wnt: A case of mistaken identity? Curr Biol 1999; 9:R717-R719.
89. Kelley LA, MacCallum RM, Sternberg MJ. Enhanced genome annotation using structural profiles in the program 3D-PSSM. J Mol Biol 2000; 299:499-520.
90. Fischer D. Hybrid fold recognition: Combining sequence derived properties with evolutionary information. Pac Symp Biocomput 2000; 5:119-130.
91. Mallick P, Goodwill KE, Fitz-Gibbon S et al. Selecting protein targets for structural genomics of pyrobaculum aerophilum: Validating automated fold assignment methods by using binary hypothesis testing. Proc Natl Acad Sci USA 2000; 97:2450-2455.
92. Rychlewski L, Jaroszewski L, Li W et al. Comparison of sequence profiles. Strategies for structural predictions using sequence information. Protein Sci 2000; 9:232-241.
93. McGuffin LJ, Bryson K, Jones DT. The PSIPRED protein structure prediction server. Bioinformatics 2000; 16:404-405.
94. Shindyalov IN, Bourne PE. Improving alignments in HM protocol with intermediate sequences. Forth Meeting on the Critical Assessment of Techniques for Protein Structure Prediction 2000; A92.
95. Gough J, Chothia C. SUPERFAMILY: HMMs representing all proteins of known structure. SCOP sequence searches, alignments and genome assignments. Nucleic Acids Res 2002; 30:268-272.
96. Novatchkova M, Eisenhaber F. A CH domain-containing N terminus in NuMA? Protein Sci 2002; 11:2281-2284.
97. Lorenz A, Wells JL, Pryce DW et al. Pombe meiotic linear elements contain proteins related to synaptonemal complex components. J Cell Sci 2004; 117:3343-3351.
98. Rabitsch KP, Gregan J, Schleiffer A et al. Two fission yeast homologs of Drosophila mei-S332 are required for chromosome segregation during meiosis I and II. Curr Biol 2004; 14:287-301.
99. Ponting CP. Issues in predicting protein function from sequence. Brief Bioinform 2001; 2:19-29.
100. Cuff JA, Clamp ME, Siddiqui AS et al. JPred: A consensus secondary structure prediction server. Bioinformatics 1998; 14:892-893.
101. Shindyalov IN, Bourne PE. Protein structure alignment by incremental combinatorial extension (CE) of the optimal path. Protein Eng 1998; 11:739-747.
102. Wildpaner M, Schneider G, Schleiffer A et al. Taxonomy workbench. Bioinformatics 2001; 17:1179-1182.
103. Devos D, Valencia A. Intrinsic errors in genome annotation. Trends Genet 2001; 17:429-431.
104. Ponting CP, Benjamin DR. A novel family of Ras-binding domains. Trends Biochem Sci 1996; 21:422-425.
105. Kalhammer G, Bahler M, Schmitz F et al. Ras-binding domains: Predicting function versus folding. FEBS Lett 1997; 414:599-602.
106. Iyer LM, Aravind L, Bork P et al. Quoderat demonstrandum? The mystery of experimental validation of apparently erroneous computational analyses of protein sequences. Genome Biol 2001; 2, (RESEARCH0051).
107. Strynadka NCJ, Eisenstein M, Katchalski-Katzir E et al. Molecular docking programs successfully predict the binding of a B-lactamase inhibitory protein to TEM-1 \BETA\-lactamase. Nature Struct Biol 1996; 3:233-239.
108. Dandekar T, Snel B, Huynen M et al. Conservation of gene order: A fingerprint of proteins that physically interact. Trends Biochem Sci 1998; 23:324-328.
109. Marcotte EM, Pellegrini M, Ng HL et al. Detecting protein function and protein-protein interactions from genome sequences. Science 1999; 285:751-753.
110. Enright AJ, Iliopoulos I, Kyrpides NC et al. Protein interaction maps for complete genomes based on gene fusion events. Nature 1999; 402:86-90.
111. Gavin AC, Bosche M, Krause R et al. Functional organization of the yeast proteome by systematic analysis of protein complexes. Nature 2002; 415:141-147.
112. von Mering C, Krause R, Snel B et al. Comparative assessment of large-scale data sets of protein-protein interactions. Nature 2002; 417:399-403.
113. Ho Y, Gruhler A, Heilbut A et al. Systematic identification of protein complexes in Saccharomyces Cerevisiae by mass spectrometry. Nature 2002; 415:180-183.
114. Schwikowski B, Uetz P, Fields S. A network of protein-protein interactions in yeast. Nat Biotechnol 2000; 18:1257-1261.

SECTION II

Complementing Biomolecular Sequence Analysis with Text Mining in Scientific Articles

CHAPTER 4

Extracting Information for Meaningful Function Inference through Text-Mining

Hong Pan, Li Zuo, Rajaraman Kanagasabai, Zhuo Zhang, Vidhu Choudhary, Bijayalaxmi Mohanty, Sin Lam Tan, S.P.T. Krishnan, Pardha Sarathi Veladandi, Archana Meka, Weng Keong Choy, Sanjay Swarup and Vladimir B. Bajic*

Abstract

One of the emerging technologies in computational biology is text-mining which includes natural language processing. This technology enables extraction of parts of relevant biological knowledge from a large volume of scientific documents in an automated fashion. We present several systems which cover different facets of text-mining biological information with applications in transcription control, metabolic pathways, and bacterial cross-species comparison. We demonstrate how this technology can efficiently support biologists and medical scientists to infer function of biological entities and save them a lot of time, paving way for more focused and detailed follow-up research.

Introduction

Text-mining of biomedical literature has received an increased attention in the past several years.[1-6] This is caused by several reasons:

a. a huge volume of the scientific documents available over internet to an average user;
b. inability of an average user to keep track of all relevant documents in a specific domain of interest;
c. inability of humans to keep track of associations usually contained in, or implied by, scientific texts; these associations could be either explicitly stated, such as 'interaction of A and B', or they need not necessarily be explicitly spelled out in a single sentence;
d. inability of humans to simultaneously deal with a large volume of terms and their cross-referencing;
f. necessity to search a number of different documents (or sometimes resources) to extract a set of relevant information;
e. inability of a single user to acquire the required information in a relatively short (acceptable) time.

As an illustration, currently PubMed repository (http://www.ncbi.nlm.nih.gov/entrez) contains over 14 million indexed documents.[1] It is common that searches of PubMed frequently provide several hundreds or more returned documents. Studying these large document sets is not an easy task for a single user. If the analysis has to be repeated several times with different selection of documents, then such a task is usually not feasible.

*Corresponding Author: Vladimir B. Bajic—Knowledge Extraction Lab, Institute for Infocomm Research, 21 Heng Mui Keng Terrace, 119613, Singapore. Email: bajicv@i2r.a-star.edu.sg

Discovering Biomolecular Mechanisms with Computational Biology, edited by Frank Eisenhaber.

Text-mining is seen as an interesting and powerful supporting technology to complement research in biology and medicine. A number of text-mining systems which tackle different problems aimed at supporting biological and medical research, and which focus on different aspects of genomics, proteomics, or relations to diseases, have been reported.[7-19]

Computational biology produces answers, which form the bases and lead to better designs for further experimentation. Among the various computational biology approaches, text-mining systems provide a unique front where large quantum of knowledge put out by experimental biologists can be efficiently screened using "vocabularies" or standard terms adopted and used widely by biologists. Hence, such systems analyze the existing knowledge and uncover potential associations among biological entities or phenomena that can lead to further experimentation. In effect, text-mining-based approaches allow biologists to focus on certain unique aspects of information that would have been reported independently thus not lending them for establishment of readily recognizable associations or correlations. Many such associations in biology go unnoticed till more directed studies are done to address the specific associations. Text mining approaches, therefore, have the inherent capacity to help speed-up the rate of biological discovery.

In this chapter we present several text-mining systems developed in our Knowledge Extraction Lab at the Institute for Infocomm Research, Singapore, two of which are the result of an on-going collaboration with Department of Biological Sciences, National University of Singapore. We show how these systems can assist an average (nonexpert) user to better understand specific problems in biology and bring them closer to the answers about functions of biological entities inferred on the basis of an in silico method. Before we present these systems, we also define the problem we intend to deal with and describe some of the general features that text-mining system should provide to the end-users.

Scope and Nature of Text-Mining in Biomedical Domain

By automated knowledge extraction, we understand an automated extraction of names of entities, such as genes, gene products, metabolites, pathways, etc., which appear in biomedical and molecular biology literature, as well as the relationships between these entities. The basic relation between two entities is characterized by the cooccurrence of their names in the same document, or in a specific segment of the document. However, the actual relation between these entities is not easy to characterize by the computer program. It is customary to leave it to the user to assess the actual nature of such relations based on the associated documents. To the best of our knowledge, very few text-mining systems exist which can accurately extract such relations.

Characteristics of Text-mining Systems

There are several basic features that text-mining systems should provide. These systems should:

a. be easy to use;
b. be interactive;
c. allow several ways of submitting data;
d. allow user to select categories of terms to be used in the analysis;
e. provide suitable interactive summary reports;
f. show association maps in suitable graphical format;
g. preferably have built-in intelligence to filter out irrelevant documents;
h. preferably be able to extract large-volume of useful information in reasonable timeframes.

While, in principle, any free text document can be analyzed, for the purposes of discussion here, we will assume that documents are abstracts of scientific articles, such as those contained in PubMed[1] repository. Then, generally speaking, there could be three levels at which text analysis can be conducted: the 'Abstract level', the 'Sentence level', the 'Relation level'. At the 'Abstract level', the system analyzes the whole abstract to determine if it contains relations between the utilized biological entities or not. At the 'Sentence level', the system assesses whether the abstract analyzed contains sentences that explicitly claim relations between the entities or not; here, the individual sentences are analyzed as a whole. Finally, at the 'Relation level', the system attempts to extract specific entities and relations they are subjected to form the sentences that are assessed to contain such relations.

The systems, which we describe here, possess different combinations of these characteristics.

Focus of Our Text-Mining Systems

Regulatory systems in different organisms perform key functions of synchronizing events in the organism at different hierarchical levels. Networks of genes, proteins and metabolites activate and operate under different intra-cellular and extra-cellular conditions in order to provide proper responses of the organism to various stimulatory signals. Understanding the cause-consequence relations between entities in such complex systems and their relations to particular pathways can help us to understand ways of how the organism functions. This can also help us to find out how to control behavior of such regulatory networks and ultimately develop more efficient drugs or reduce problems of genetic-based diseases. Text-mining systems that we have developed aim at helping individual researchers to elucidate and partially reconstruct segments of such networks. Our text-mining systems focus on three general domains: transcription control, metabolic networks and gene networks. These are explained in more detail below.

Supporting Text-Mining Systems for Gene Regulatory Networks Reconstruction

Transcription control is the key mechanism for activation of genes and gene groups under different cellular conditions. Transcriptional regulatory networks[20] provide information necessary to study different modalities of gene activation, such as tissue-specificity, timing, and rate of gene transcription. To infer parts of these control networks, we need to know the relations between different transcription factors (TFs) involved in the process of transcription initiation. For example, let us assume that hypothetical TF A and TF B are both gene products of genes GA and GB, respectively, and that a gene G requires a complex formed by A and B to bind to its promoter in order that transcription of G in a particular tissue can be activated. Then, if in the nucleus of the target cells there is no complex A/B, gene G will not activate. The necessary condition obviously is that both genes GA and GB produce A and B in sufficient quantities so that A/B complex can be generated. Thus, activation of G requires activity of GA and GB. Also, in order that GA and GB be activated another set of TF would be required, and so on. In this way, one can step by step reconstruct parts of the necessary gene networks which induce activation of gene G. However, we also notice that due to interaction of A and B a synchronous action of GA and GB would be expected, meaning that such two genes should be able to coexpress under particular specific conditions in some tissues. This, again, suggests that their promoters should share some degree of similarity in terms of promoter context (the type of TF binding sites, their ordering and partly spacing).[21] Thus analysis of promoters of one of these genes, say GA, may reveal part of the information about the promoters of GB. As can be seen in this hypothetical, but very common case, information about relation between A and B (they form a complex A/B) and information that A/B binds to promoter of G, provides many additional clues what to look for in reconstructing gene network which control activation of G in the target tissue. Text-mining tools can assist us tremendously in such tasks.

Relations between TFs are thus one of the key sources of information for reconstruction of parts of gene transcriptional regulatory networks. The TRANSCompel database[22] is a resource which contains several hundreds experimentally verified relations between TFs collected manually from the literature. These relations need not always be of the form of interactions. For example, it is possible to have information of the form: 'if both present, A and B affect transcription of G'. These forms of cause-consequence relations may generally show synergistic effect on G when they enhance G's transcription, or antagonistic effect when they negatively impact transcription of G. Unfortunately, the content of TRANSCompel database

Transcription Factor Interaction Extraction Result

TransFactor 1	Relation Word	TransFactor 2	PMID	Sentence
Sp1	bind	Sp1	14518567	The transcription factor **Sp1** binding to the two **Sp1** recognition motifs residing at -158 to 150 bp and -123 to 114 bp in the gene promoter is found to be essential for both EGF- and PMA-induced gene expression of human 12(S)-lipoxygenase
c-Jun	interact	Sp1	14518567	Since both of the transcription factors **c-Jun** and **Sp1** are prerequisite for EGF and PMA response, interaction between **c-Jun** and **Sp1** may account for the functional regulation of human 12(S)-lipoxygenase gene regulation
c-Jun	cooperate	Sp1	14518567	The direct and cooperative interaction between **c-Jun** and **Sp1** induced by EGF or PMA activates the expression of the human 12(S)-lipoxygenase gene
c-Jun	interact	Sp1		
ER alpha	interact	Stat3	12709435	Based on the data presented, we hypothesized that in the presence of prothymosin alpha, **ER alpha** activates its direct target genes and increases cell proliferation, whereas in the presence of high levels of TGF-alpha, **ER alpha** preferentially interacts with **Stat3** and causes cell differentiation

Figure 1. A snapshot of the report page generated in the analysis of two PubMed documents is shown. First and third columns show the names of TFs, the second column shows the relation word, the fourth column gives a link to the PubMed document from which information is extracted. The column 'Sentence' shows the sentence from which the relation has been extracted with important words highlighted.

represents only a fraction of TF relations which is documented. It is a great challenge to collect this information and make it available to researchers in gene regulatory networks field. One of our systems, Dragon TF Relation Extractor, allows such direct extraction of actual relations.

Dragon TF Relation Extractor (DTFRE)

This section describes DTFRE, a system that extracts the actual TF names and the type of relation(s) between them. The system is available for public and nonprofit use at http://research.i2r.a-star.edu.sg/DRAGON/TFRE/. DTFRE is developed based on manually cleaned large corpus of data. Within 8 months, five trained biologists and a chemist have read, analyzed and classified more than 3000 PubMed abstracts related to transcription control in eukaryotes. Based on this information we have generated a database of classified sentences about TF relations. This database (TFRD) contains 5244 sentences. Out of these, 2402 sentences have been classified positive, i.e., as those which contain explicit statements about TF relations; 1766 sentences have been classified as negative entries, while there have been 1076 entries classified as ambiguous. To the best of our knowledge, TFRD represent the largest, manually cleaned dataset to date used for developing specialized text-mining systems for extracting TF relations from biomedical literature.

The system allows user to submit data and to obtain report which lists individual TFs and the extracted relations between them, as detected from the supplied data. Data to be submitted are simply the PubMed abstracts obtained as results of the search of PubMed database by the Entrez system and saved or copied in the text form. Generated reports are interactive. To simplify analysis of extracted relations for users, we provided visualization of the actual sentence from which the relation is extracted, as well as a link to PubMed original document and to such a document with marked crucial words and sentences. Users have to provide e-mail address to which the link to their result files will be given. In Figure 1, we give a snapshot of a report generated from the analysis of PubMed documents ID = 14518567 and ID = 12709435, while in Figure 2, we show the first document with colored segments, as explained above.

PMID : 14518567

TITLE : Cell signaling and gene regulation of human 12(S)-lipoxygenase expression.

ABSTRACT : Human 12(S)-lipoxygenase is a platelet-type 12(S)-lipoxyenase. Its expression is detected in human erythroleukemia cells, human skin epidermal cells and human epidermoid carcinoma A431 cells. Treatment of A431 cells with EGF or PMA induces the gene expression of human 12(S)-lipoxygenase. The induction of gene expression is mediated through the cell signaling of MAPK activation, followed by the induction of **c-Jun** expression. The transcription factor **Sp1** binding to the two **Sp1** recognition motifs residing at -158 to 150 bp and -123 to 114 bp in the gene promoter is found to be essential for both EGF- and PMA-induced gene expression of human 12(S)-lipoxygenase. However, no change of **Sp1** binding to GC-rich sequence was observed while no **AP-1**-binding site can be found in the responsive region of the promoter in EGF- and PMA-induced promoter activation of the human 12(S)-lipoxygenase gene. Since both of the transcription factors **c-Jun** and **Sp1** are prerequisite for EGF and PMA response, interaction between **c-Jun** and **Sp1** may account for the functional regulation of human 12(S)-lipoxygenase gene regulation. The direct and cooperative interaction between **c-Jun** and **Sp1** induced by EGF or PMA activates the expression of the human 12(S)-lipoxygenase gene. Therefore, **Sp1** may serve at least in part as a carrier to bring **c-Jun** to the promoter, thu's transactivating the transcriptional activity of the human 12(S)-lipoxygenase gene..

Transcription Factor List

c-Jun
AP-1
Sp1

Figure 2. An example of a colored PubMed document with identified TF names highlighted in red, with marked sentences from which TF relations are extracted, and with relation words highlighted in blue. At the bottom of the page the list of identified TF names in the analyzed document are shown. A color version of this figure is available online at http://www.Eurekah.com.

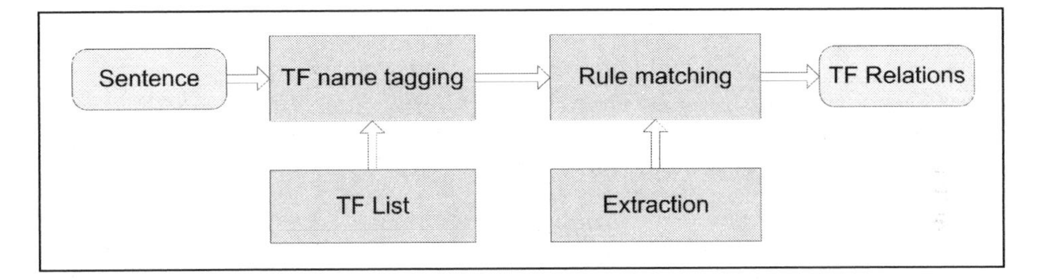

Figure 3. Conceptual structure of DTFRE.

DTFRE is a rule-based system. Given a sentence, the system first tags TF names in the input sentences using a prebuilt TF dictionary. Then the sentences are matched with the rules in a rulebase. For every match, slots containing TFs and relation are extracted and presented to the user (Fig. 3).

Rulebase Construction

A crucial component of the system is the rulebase that captures the knowledge about the TF-TF relation patterns in the sentences. Construction of the rulebase is a nontrivial task. Traditional method is to hand-code the rules with the help of experts.[23-25] However, hand-coding is a tedious task and is error-prone, especially when there are a lot of abstracts to be analyzed. Automatically learning the rules from abstracts is an attractive alternative. Learning extraction rules could help automate the rulebase construction or at least ease the hand-coding process, e.g., by letting the learning method generate seed rules that could be manually refined.

Rule learning from text is an active topic investigated by the Information Extraction (IE) community.[26] Though a number of rule learning systems have been proposed,[27] directly

```
Inf(interact) – 2 – between – 15 – TF – 2 – and – 2 – TF
V
TF – 15 – Inf(interact) – 4 – with – 8 – TF
V
...
```

Figure 4. Sample rules for 'interact' relation.

applying them to extract biological interactions has produces only moderate results.[28] The reason is that the complexities in biomedical literature demand learning algorithms customized for the biomedical domain. We have developed one such learning algorithm for constructing the rulebase of our system.

Rule Representation

Our rule learning algorithm uses a disjunctive rule representation. An example of the rule for the "interact" relation is shown in Figure 4. As observed in the figure, the rule consists of several regular patterns connected by the disjunction operator. The regular patterns follow a specific format as below. Every pattern:

- Has exactly two TFs and one relation word (possibly an inflexion).
- Has connector words (optional).
- Has intra term distance limits.

For example, consider the first regular pattern in Figure 4. It has two TF names and an inflexion of the relation word "interact". The connector word is "with" which appears between the relation word and the second TF name. The distances between the adjacent terms stand for the maximum number of wildcard words that could be tolerated between the respective terms, for a rule match.

Learning Algorithm

Figure 5 presents an outline of our algorithm for learning the disjunctive rules from a training sentence corpus. The algorithm picks a random positive example and attempts to identify a candidate rule pattern that has support and confidence above minimum specified values. All positive examples covered by this pattern are removed and more candidate rules are generated until all positive examples are covered. The candidate list is then pruned to remove insignificant rules and the remaining rules represent the learned rules. The algorithm is run for each relation separately and hence there is at least one rule for every relation. The rulebase is simply a collection of all the learned rules.

The learned rules were evaluated using 3-fold cross validation. We obtained over 90% precision and 75% recall on average for the seven types of relation words ('interact', 'complex', 'bind', 'associate', 'synergise', 'cooperate', 'inhibit') we used in this system. We spent about 300 man hours to manually tune the learned rules and obtained significant increase in accuracy reflected in 93% precision and 88% recall. For comparison, SUISEKI's[14] reported performance in extraction of protein-protein interactions is 46% precision and 40% recall, while PreBIND[16] performed at 92% precision and 92% recall, but only for a restricted problem of classifying sentences as describing protein-protein interactions or not. PreBIND does not extract the actual relations.

DTFRE is the first public system for TF relation extraction. It achieves accuracy characterized by 93% precision and 88% recall on our test data, which is very similar in performance to that of single-pass manual curation. However, it will be dangerous to extrapolate this performance to an

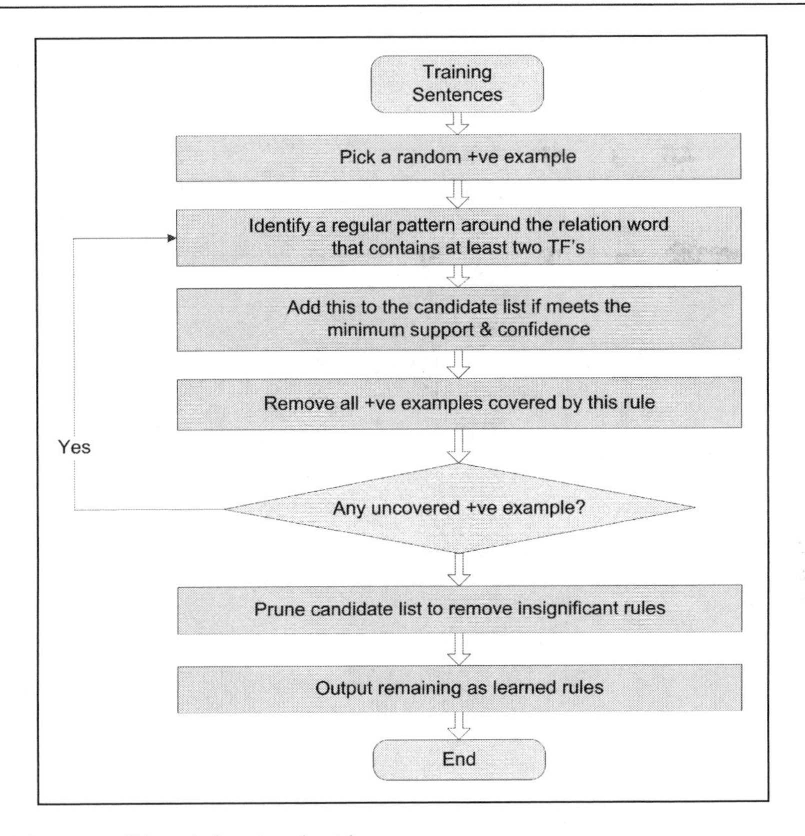

Figure 5. Structure of the rule-learning algorithm.

arbitrary set of documents, since the volume of the data used in training and testing is still very small (although it is the largest corpus of manually curated data used for similar tasks in biomedical text-mining). The system is based on a combination of automatic learning for the generation of extraction rules and manual rule tuning. The learning method uses a representation that is human comprehensible and hence the learned rules are easy to manually verify and tune to achieve best performance. With the rule learning algorithm, we were able to cut down the hand-crafting time considerably. However, the rule representation is shallow and cannot accurately recognize relations expressed in complex sentence structures, e.g., through coreferences. We are addressing this issue as part of our current work.

Mining Associations of Transcription Factors by Dragon TF Association Miner

While DTFRE aims at identifying and actually extracting specific relations and TFs subjected to such relations, the goal of Dragon TF Association Miner (DTFAM) is different. It aims at providing more broad information about potential association of TFs with concepts from Gene Ontology (GO),[29] as well as with diseases, in order to help biologists and medical researchers to infer unusual functional associations. The system uses five well-controlled vocabularies. Three vocabularies are related to GO (biological process, molecular function, and cellular component), while the fourth one is related to different disease states. The fifth vocabulary contains TF names. Functional associations of TFs to any term from the four categories (GO and diseases) can be focused to any combination of these terms, such as biological

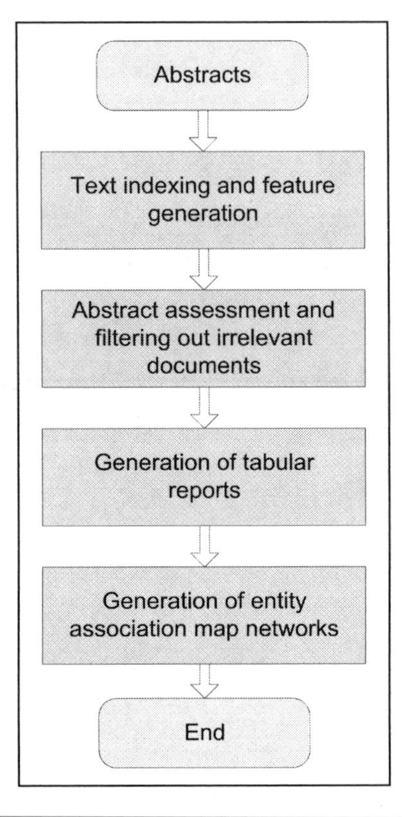

Figure 6. Schematic presentations of DTFAM structural modules

process, or biological process and diseases, etc., depending on the user's selection. All GO vocabularies are general. Disease vocabulary is focused to human diseases, while the TF vocabulary contains over 10,000 TF names and their synonyms collected for various species, mainly eukaryotes, but also including *E. coli*, *B. subtilis* and some other prokaryotes. Some necessary data cleaning has been done with all vocabularies in order to enable more efficient text-mining.

DTFAM can be accessed freely for academic and nonprofit users at http://research.i2r.a-star.edu.sg/DRAGON/TFM/. The system is trained and tested on the previously described corpus of 3000 PubMed documents which were manually classified. The system attempts to assess at the 'Abstract level' whether the document analyzed contains information about TF relations or not. The user has possibility to select the level of filtering out irrelevant documents. This function reduces the 'noise' (i.e., usage of irrelevant documents) considerably for the generation of final reports. However, it cannot eliminate the irrelevant documents completely.

There are several modules which operate within the system (see Fig. 6):

- The first module analyses the submitted text, makes necessary indexing of terms and generates features for the intelligent module.
- The second module analyses the content of the processed document and applies one of 65 previously derived models in assessing whether the analyzed document should be retained or rejected. If the model signals that the document contains information about TF relations, the document is accepted for the final analysis, otherwise it is rejected. The selection of models is automatic and it is determined by the selected sensitivity on the systems main page. The higher the sensitivity, the more documents will be selected for the analysis, but

this may also include a large number of irrelevant documents. These 65 models are developed and tuned based on specific feature selection, signal processing, nonlinear modeling, artificial neural networks and discriminant analyses.

- The third module generates interactive tabular reports.
- The fourth module analyses the connections (associations) between the terms and generates interactive association map networks of these terms. The association of terms is based on their cooccurrence in the same PubMed document. The nodes of the generated graphs represent the terms from the selected vocabularies. Different shapes and coloring is used to make it easy for users to analyze these graphs. All nodes are interactive and by clicking on the node a set of related PubMed documents with color-marked terms will be opened for user's inspection and assessment of the relevance of proposed associations.

The main characteristics of DTFAM system are:

1. It is focused on exploring potential association of TFs with other important functional categories such as GO terms and diseases.
2. It provides suitable interactive reports both tabular and graphical.
3. Its module for filtering irrelevant document has been trained on a unique large manually-curated corpus of data.
4. It uses five manually-curated vocabularies (one for TF names and synonyms, three for GO categories, one for diseases).

This system is unique in the combinations of features and utility it provides to the users. The most distinctive features are its focus on transcriptional regulation, its module for filtering out irrelevant documents trained on manually-curated large data corpus related to relations between TFs, and possibility for the user to select the stringency of filtering irrelevant documents.

To illustrate how, in a simple way, users can extract useful information by this system, we will assume that it is of interest to find out what are TFs potentially involved in the toll-like receptor mediated activation of signaling pathway which induces an antimicrobial innate immune reaction.[30] We also want to find out what are the biological processes, molecular functions, cellular components and diseases that could be associated with the found TFs. Antimicrobial peptides are constitutive ingredients of innate immunity and they take role of the first layer of defense of the host against invading pathogens. Some of these peptides are gene products and can be transcriptionally activated. For example, in *Drosophila*, Toll signaling pathway regulates rapid production of antimicrobial peptides in response to infection by pathogens. We will perform this exploration by selecting a query 'toll antimicrobial'. We will also select Sensitivity of 0.95, and all four vocabularies at the main page. As a part of the analysis and reports, the system will generate two association map networks. Analysis of the first network depicted in Figure 7, reveals that DFTAM detected inhibitor kappaB (IkappaB), NF-kappaB and c-Jun as TFs relevant for this signaling pathway. The roles of these three TFs in this pathway are documented in.[30-32] All other entities found and presented in the network relate to proper GO categories, immune response and diseases. We also observe that *Drosophila* TFs, Cactus and Dorsal, have been found. Cactus is IkappaB-like TF, while Dorsal is NF-kappaB-like TF. This shows that DTFAM is capable of extracting relevant biological knowledge. We suggest, however, that a user should not blindly accept results of the analysis and should evaluate the relevance of detected associations by consulting the references used by the system. Since the system provides interactive graphs with links to the documents used, as well as color-highlighted the terms used in the analysis, this task is made easier for the user.

Exploring Metabolome of *Arabidopsis thaliana* and Other Plant Species by Dragon Metabolome Explorer

The largest category of gene functions in all the eukaryotic genomes sequenced thus far is that of metabolism, which can comprise almost 25% of all genes.[33] Metabolome, in its complete sense, includes all metabolic pathways and their components, including the enzymes

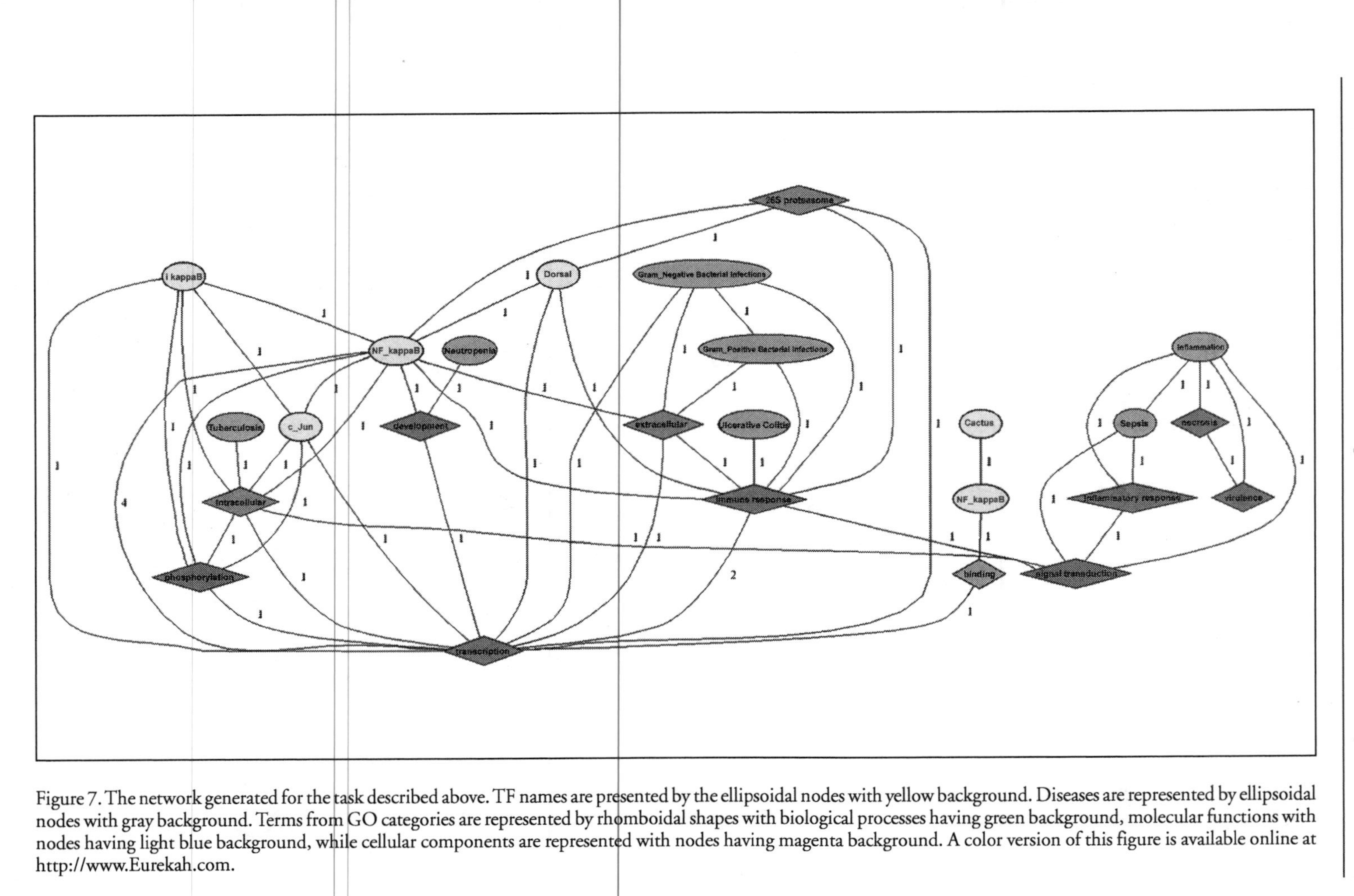

Figure 7. The network generated for the task described above. TF names are presented by the ellipsoidal nodes with yellow background. Diseases are represented by ellipsoidal nodes with gray background. Terms from GO categories are represented by rhomboidal shapes with biological processes having green background, molecular functions with nodes having light blue background, while cellular components are represented with nodes having magenta background. A color version of this figure is available online at http://www.Eurekah.com.

and the regulators. In this section we present a system, Dragon Metabolome Explorer (DME), for the exploration of metabolic subsystems in plants and their associations with genes and all the GO categories summarized in ontologies adopted by the Arabidopsis research community (www.tair.org). In addition to general GO categories, such as biological processes, molecular functions, and cellular components, Arabidopsis specific ontologies which DME uses are related to anatomic parts and developmental stages. The exploratory analysis of associations of the GO terms/entities can suggest meaningful functional links and pave way for a more detailed and focused analysis using experimental approaches. The system is free for academic and nonprofit users and can be accessed at http://research.i2r.a-star.edu.sg/DRAGON/ME_v2/.

Metabolic processes control body functions through highly complex networked pathways. Many small molecules associated with such pathways act as regulators of genes and diverse cellular functions. Understanding metabolic processes, therefore, is one of the key issues of modern biology and requires a systems approach due to their complex nature. The limits of metabolic complexity are found in plants due to their extensive secondary metabolism networks; therefore, they form excellent resources for developing knowledge extraction tools which target metabolic pathways. The model plant, *Arabidopsis thaliana*, has 185 metabolic pathways documented, including over 700 different compounds and nearly 525 enzymes.[34,35] However, the information about pathways is not complete, which explains the fact that there are approximately 900 known metabolites found experimentally in *Arabidopsis*, which are not assigned to any of the known metabolic pathways.

The currently available public resources on metabolic pathways are almost exclusively devoted towards representing the metabolic pathways. Some of these are, MetaCyc,[36] ENZYME,[37] BRENDA,[38] IntEnz,[39] KEGG,[40] PathDB [http://www.ncgr.org/research/pathdb/], UM-BBD.[41] WIT2.[42] Some of the above recent resources such as AraCyc additionally link the pathway information to the genome resources.[35] However, to the best of our knowledge, there are currently no resources for extracting knowledge of the function of metabolites and pathways from the existing literature with the aim to complement the pathway related information. Our system, DME, is one such bioinformatic tool that can support research in this direction and can simplify task for individual biologists. It has sufficient flexibility and provides comprehensive summarized information in a form suitable for simple use by biologists. The information provided is also with the high coverage, attempting to include much of the known knowledge.

The algorithm is based on text analysis of PubMed documents. The system uses several highly controlled vocabularies and matches cooccurrence of terms from these dictionaries within a set of documents, and determines significance of each of these terms. It provides users comprehensive listings of three categories of metabolome components found in the analyzed documents—pathways, enzymes, metabolites, and any of the categories from the additional three vocabularies specific for *Arabidopsis thaliana* (related to anatomy, developmental stages, genes), as well as those related to cellular component, biological process, and molecular function. DME attempts to detect potential associations between the terms form these vocabularies and produces different reports including networks of associations. All reports including graphical ones are interactive and contain hyperlinked nodes to provide PubMed abstracts directly.

There are three possible ways a user can submit documents for the analysis. Documents can be selected by forming any query acceptable for Entrez search engine of PubMed repository, or the user can perform PubMed search in advance and save selected documents in the text format, and then submit such saved documents to the program for the analysis. The second mode is preferable. The system possesses great flexibility as it allows arbitrary query to be submitted for the abstract selection. The tabular report presents every term from the selected vocabularies found in the document set and links of the PubMed documents where the terms have been found. These terms are also provided in three colors (green, red and blue for pathways, metabolites and enzymes, respectively) for easier visual inspection. Graphical report may contain several association networks that depict the terms found in the analysis. Each term is represented by an oval node (green nodes denoting pathways, yellow node with

List of single names.

Please note, names in green color are from PATHWAY, in red color are from COMPOUND, and in blue color are from ENZYME. and in orange color are from ANATOMY.

PATHWAY	COMPOUND	ENZYME	ANATOMY	Frequency	PubMed IDs
	2-oxoglutarate			3	11208024; 12696919; 9863248;
		2-oxoglutarate-dependent dioxygenase		1	12782296;
		CATALASE		1	9863248;
		anthocyanidin synthase		1	14570878;
anthocyanin biosynthesis				2	10417729; 12126705;
	ascorbate			2	11208024; 9863248;
	carbon monoxide			1	8507822;
	catechin			1	14976232;
		chalcone isomerase		2	12620339; 8507822;
		cinnamate 4-hydroxylase		1	8507822;
	cyanide			1	8507822;
		dihydroflavonol 4-reductase		4	10417729; 11724379; 12126705; 12955207;
	dihydrokaempferol			21	10417729; 11409967; 11596309; 11724379; 12018025; 12126705; 12180990; 12620317; 12620339; 12643678; 12696918; 12696919; 12711137; 12782296; 12955207; 13921745; 14609128; 14976232; 2221917; 8507822; 9863248;
	dihydroquercetin			7	10417729; 12018025; 12180990; 12620339; 12696919; 12711137; 14976232;
	flavanone			4	11208024; 12643678; 12782296; 14570878;
		flavonoid 3'-hydroxylase		3	12018025; 12620339; 8507822;
		flavonol synthase		4	12696918; 12696919; 12782296; 14570878;
			flower	5	10417729; 11724379; 12018025; 12126705; 12955207;
			inflorescence	1	11409967;

Figure 8. Part of the interactive tabular report of DME using the term "dihydrokaempferol".

red letters denoting metabolites, and blue nodes denoting enzymes). Again, different colors help easier inspection of the generated association networks.

We illustrate here how this system can be used to infer function related to metabolic pathways. To do so, we take the example of the pathways, metabolites, enzymes and plant anatomy terms associated with the activity of the metabolite, dihydrokaempferol. The query was dihydrokaempferol. The selected vocabularies were: pathways, metabolites, enzymes and anatomy. System produced an interactive tabular report, part of which is shown in Figure 8, and an interactive association map network depicted in Figure 9.

DME found anthocyanin biosynthesis pathway of which dihydrokaempferol is part. It also found 18 compounds, 9 enzymes and 8 anatomy parts where this metabolite is found. From components of the anthocyanin biosynthesis pathway, DME identified 5 out of 8 metabolites and 2 out of 3 enzymes. DME also displayed 5 out of 10 metabolites and 3 out of 9 enzymes which are present in flavonoid biosynthesis pathway. Some of these metabolites and enzymes are shared between these two pathways, suggesting links between anthocyanin biosynthesis and flavonoid biosynthesis pathways. These two pathways fall under more general phenylpropanoid pathway. Also, for example, DME has found that flower is related to dihydroflavonol 4-reductase, flavonoid 3'-hydroxylase, dihydrokaempferol, leucopelargonidin, anthocyanin biosynthesis. It is, however, documented that anthocyanin biosynthesis pathway is involved in flower pigmentation.[43] Additionally, the above mentioned enzymes and

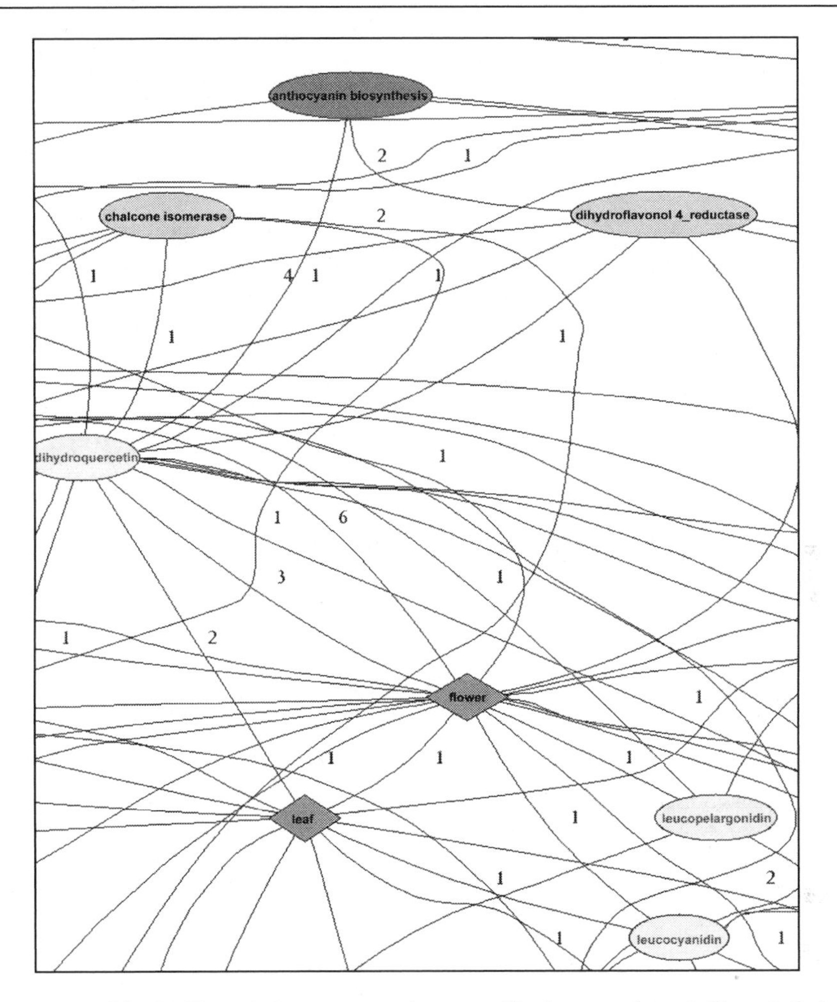

Figure 9. A part of the DME association map network generated by the query shown in Figure 8. Pathways, enzymes, metabolites and plant anatomy terms are shown in different shaped and colored nodes.

metabolites from anthocyanin biosynthesis are involved in flower pigmentation (dihydroflavonol 4-reductase does not show pigmentation of flower due to the accumulation of dihydrokaempferol;[43] flavonoid 3'-hydroxylase shows pigmentation of anthocyanins;[44] leucopelargonidin gives the orange color to flowers). Moreover, Dihydroflavonol 4-reductase inefficiently reduces dihydrokaempferol in anthocyanin biosynthesis[45] and DME linked these together. These several extracts from the reports of DME are used to illustrate that one can infer many specific issues related to function of metabolic subsystems.

Comparative Analysis of Bacterial Species

One of the interesting possibilities is the use of text-mining in the cross-species studies. The aim of such tasks is to find out in an automated fashion the facts common to two or more species, as well as those specific for individual species or group of species. For example, we may be interested in finding common parts of complex regulatory networks and pathways which are preserved in various species (and thus common), as well as to find out gene networks characteristic of separate species development related to the same or similar pathways. Due to the putative nature of text-mining, this approach is highly useful in suggesting functional associations between the entities searched in a given framework.

Our system, Dragon Explorer of Bacterial Genomes (DEBG), has currently data for two bacteria, *Pseudomonas aeruginosa* and *Escherichia coli*. We plan to extend it to other microbes shortly. DEBG contains species-specific vocabularies of genes (and their synonyms), each containing several thousand entries. The system analyses cooccurrence of terms from these vocabularies in several groups of carefully selected documents from PubMed repository, and summarizes results obtained. Then it provides interactive graphical and tabular presentations of the associations found. DEBG relies on a local installation of PubMed.

Users can supply up to three concepts to be used in the selection of documents. One of the concepts should be broad, while the other two should be more specific. DEBG will automatically form several queries to collect documents required for the analysis. We will illustrate this through a particular example.

To illustrate the use of DEBG, let us assume that we are interested in exploring the differences in gene networks controlling flagellar motility and twitching motility[46-48] in our two bacterial species. We can select 'motility' as the broad category, while 'flagella' and 'twitching OR fimbriae OR pili' can be selected for the more specific categories. The reason we added 'fimbriae' and 'pili' is because twitching motility in *E. coli* is commonly associated with fimbriae, while in *P. aeruginosa*, it is the type VI pili. Example of queries which DEBG forms are as follows:

Q1: (Pseudomonas OR '*Escherichia coli*') AND flagella

Q2: (Pseudomonas OR '*Escherichia coli*') AND (twitching OR fimbriae OR pili)

Q3: (Pseudomonas OR '*Escherichia coli*') AND motility

After the documents are collected the system identifies the existing terms from the vocabularies used, index all found terms as belonging to one or another organism, or both, and also the category (motility, flagella, twitching). Based on these summary results, the system generates interactive tabular and graphical reports.

The system provides a colored output for the different groups of genes found in documents specific to queries Q1 and Q2. In the gene association networks, the nodes represent the genes. Genes specific to one species are shown with one shape of nodes, while those form the other species with different shape of nodes. Genes found in the documents in response to query Q3 are also depicted as a separate shaped nodes and in different color. This different coloring and shapes of nodes related to categories and species make inspection and analysis of the found networks much easier for biologists.

For example, in Figure 10, one may observe that genes in generated network appear in three big groups, one yellow-colored corresponding to twitching motility, the other magenta-colored corresponding to flagellar motility, and the third one green colored corresponding to genes contained in documents related to 'motility', but not directly related to flagellar or twitching motility. One can easily track the association of genes to species, as well as potential associations of other found genes supposedly involved in motility, but not necessarily associated with the two specific types of movement. An interesting observation is that *fliC* (Fig. 10 bottom panel), the most abundant structural component of the flagellar apparatus, is linked extensively to other genes in the network, indicating its importance in the formation of the flagella. Also, genes with related functions are likely to be located in close proximity, for example, the *che* genes involved in chemotaxis are clustered together in the network.

The conclusions of this in silico experiment, which included 3522 documents in total, are that in a relatively simple fashion and in a short time, we are capable to summarize a part of information regarding to these two types of movements in two bacteria and obtain rich material for further detailed analysis.

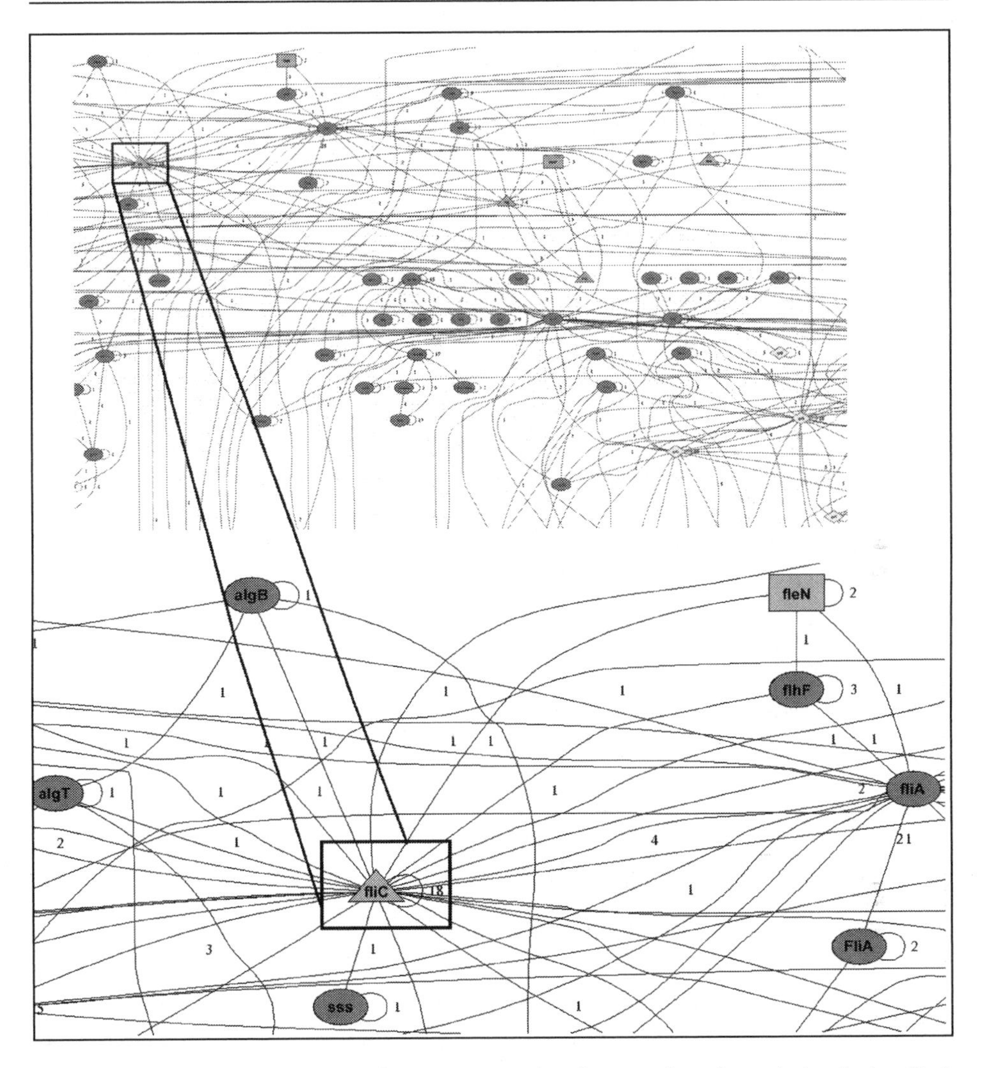

Figure 10. A part of a complex network of gene associations based on textual searches and related to 'motility', 'flagellar motility' and 'twitching motility' in *P. aeruginosa* and *E. coli*. The flagellar structural protein *FliC* and its associations with some other flagellar structures is shown in the close-up panel at bottom.

Conclusions

We show here that text-mining is a useful technology that can support research in life-sciences and allow easier inferences of function of examined entities. The strength of this approach is its comprehensiveness and ability to present sometimes unexpected associations of categories and terms based on analysis of large sets of documents. This is not feasible for a single user. However, this also is a weakness, since very few text-mining systems have built-in intelligence to automatically determine the relevance from the document context. The accuracy of such intelligent blocks is currently not sufficiently high, which requires that users carefully analyze the results obtained. However, the developments of natural language processing will make crucial contributions to this growing field in the future.[49]

References

1. Wheeler DL, Church DM, Edgar R et al. Database resources of the National Center for Biotechnology Information: Update. Nucleic Acids Res 2004; 32:D35-40.
2. Dickman S. Tough Mining: The challenges of searching the scientific literature. PLoS Biol 2003; 1(2):E48.
3. de Bruijn B, Martin J. Getting to the (c)ore of knowledge: Mining biomedical literature. Int J Med Inf 2002; 67(1-3):7-18.
4. Grivell L. Mining the bibliome: Searching for a needle in a haystack? New computing tools are needed to effectively scan the growing amount of scientific literature for useful information. EMBO Rep 2003; 3(3):200-203.
5. Andrade MA, Bork P. Automated extraction of information in molecular biology. FEBS Lett 2000; 476(1-2):12-17.
6. Schulze-Kremer S. Ontologies for molecular biology and bioinformatics. In Silico Biol 2002; 2(3):179-193.
7. Jenssen TK, Laegreid A, Komorowski J et al. A literature network of human genes for high-throughput analysis of gene expression. Nat Genet 2001; 28(1):21-28.
8. Tanabe L, Scherf U, Smith LH et al. An Internet text-mining tool for biomedical information, with application to gene expression profiling. Biotechniques 1999; 27(6):1210-4, (1216-7).
9. Perez-Iratxeta C, Perez AJ, Bork P et al. Update on XplorMed: A web server for exploring scientific literature. Nucleic Acids Res 2003; 31(13):3866-3868.
10. Becker KG, Hosack DA, Dennis Jr G et al. PubMatrix: A tool for multiplex literature mining. BMC Bioinformatics 2003; 4(1):61.
11. Asher B. Decision analytics software solutions for proteomics analysis. J Mol Graph Model 2000; 18:79-82.
12. Hosack DA, Dennis G, Sherman BT et al. Identifying biological themes within lists of genes with EASE. Genome Biology 2003; 4:R70.
13. Kim SK, Lund J, Kiraly M et al. A gene expression map for Caenorhabditis elegans. Science 2001; 293:2087-2092.
14. Blaschke C, Valencia A. The potential use of SUISEKI as a protein interaction discovery tool. Genome Inform Ser Workshop Genome Inform 2001; 12:123-34.
15. Chiang JH, Yu HC, Hsu HJ. GIS: A biomedical text-mining system for gene information discovery. Bioinformatics 2004; 20(1):120-121.
16. Donaldson I, Martin J, de Bruijn B et al. PreBIND and Textomy—mining the biomedical literature for protein-protein interactions using a support vector machine. BMC Bioinformatics 2003; 4(1):11.
17. Perez-Iratxeta C, Bork P, Andrade MA. Association of genes to genetically inherited diseases using data mining. Nature Genetics 2002; 31:316-319.
18. Chiang JH, Yu HC. MeKE: Discovering the functions of gene products from biomedical literature via sentence alignment. Bioinformatics 2003; 19(11):1417-1422.
19. Srinivasan P. MeSHmap: A text mining tool for MEDLINE. Proc AMIA Symp 2001; 642-646.
20. Lee TI, Rinaldi NJ, Robert F et al. Transcriptional regulatory networks in saccharomyces cerevisiae. Science 2002; 298:799-804.
21. Werner T, Fessele S, Maier H et al. Computer modeling of promoter organization as a tool to study transcriptional co regulation. FASEB J 2003; 17(10):1228-37.
22. Kel-Margoulis OV, Kel AE, Reuter I et al. A database on composite regulatory elements in eukaryotic genes. Nucleic Acids Res 2002; 30(1):332-4.
23. Thomas J, Milward D, Ouzounis C et al. Automatic extraction of protein interactions from scientific abstracts. Pacific Symposium on Biocomputing 2000; 5:538-549.
24. Blaschke C, Valencia A. The frame-based module of the Suiseki information extraction system. IEEE Intelligent Systems 2002; 17:14-20.
25. Ono T, Hishigaki H, Tanigami A et al. Automated extraction of information on protein-protein interactions from the biological literature. Bioinformatics 2001; 17(2):155-161.
26. Appelt DE, Israel D. Introduction to information, extraction technology. Proc of International Joint Conference on Artificial Intelligence (IJCAI-99), Stockholm, Sweden: 1999, (URL: http://www.ai.sri.com/~appelt/ie-tutorial/).
27. Muslea I. Extracting patterns for information extraction tasks: A survey. The AAAI Workshop on Machine Learning for Information Extraction 1999, (URL: http://www.ai.sri.com/~muslea/papers.html).

28. Bunescu R, Ge RF, Kate RJ et al. Learning to extract proteins and their interactions from medline abstracts. Proceedings of the ICML-2003 Workshop on Machine Learning in Bioinformatics 2003; 46-53.
29. Harris MA, Clark J, Ireland A et al. Gene ontology consortium. The Gene Ontology (GO) database and informatics resource. Nucleic Acids Res 2004; 32:D258-61.
30. Telepnev M, Golovliov I, Grundstrom T et al. Francisella tularensis inhibits Toll-like receptor-mediated activation of intracellular signaling and secretion of TNF-alpha and IL-1 from murine macrophages. Cell Microbiol 2003; 5(1):41-51.
31. Takeuchi O, Akira S. Toll-like receptors; their physiological role and signal transduction system. Int Immunopharmacol 2001; 1(4):625-35.
32. Lee SJ, Lee S. Toll-like receptors and inflammation in the CNS. Curr Drug Targets Inflamm Allergy 2002; 1(2):181-91.
33. The arabidopsis genome initiative, analysis of the genome sequence of the flowering plant arabidopsis thaliana. Nature 2000; 408:796.
34. Mueller. AraCyc: A biochemical pathway database for arabidopsis. Plant Physiol 2003; 132:453-460.
35. Rhee SYl. The Arabidopsis Information Resource (TAIR): A model organism database providing a centralized, curated gateway to arabidopsis biology, research materials and community. Nucleic Acids Res 2003; 31:224-228.
36. Krieger CJ, Zhang P, Mueller LA et al. MetaCyc: A multiorganism database of metabolic pathways and enzymes. Nucleic Acids Res 2004; 32:D438-442.
37. Bairoch A. The ENZYME database in 2000. Nucleic Acids Res 2000; 28:304-305.
38. Pharkya P, Nikolaev EV, Maranas CD. Review of the BRENDA database. Metab Eng 2003; 5(2):71-3.
39. Fleischmann A, Darsow M, Degtyarenko K et al. IntEnz, the integrated relational enzyme database. Nucleic Acids Res 2004; 32:D434-7.
40. Kanehisa M, Goto S, Kawashima S et al. The KEGG resource for deciphering the genome. Nucleic Acids Res 2004; 32:D277-80.
41. Ellis LB, Hershberger CD, Bryan EM et al. The university of minnesota biocatalysis/biodegradation database: Emphasizing enzymes. Nucleic Acids Res 2001; 29(1):340-3.
42. D'Souza M, Romine MF, Maltsev N. SENTRA, a database of signal transduction proteins. Nucleic Acids Res 2000; 28(1):335-6.
43. Johnson ET, Yi H, Shin B et al. Cymbidium hybrida dihydroflavonol 4-reductase does not efficiently reduce dihydrokaempferol to produce orange pelargonidin-type anthocyanins. Plant J 1999; 19(1):81-5.
44. Owens DK, Hale T, Wilson LJ et al. Quantification of the production of dihydrokaempferol by flavanone 3-hydroxytransferase using capillary electrophoresis. Phytochem Anal 2002; 13(2):69-74.
45. Prescott AG, Stamford NP, Wheeler G et al. In vitro properties of a recombinant flavonol synthase from arabidopsis thaliana. Photochemistry 2002; 60(6):589-93.
46. Macnab RM. How bacteria assemble flagella. Annu Rev Microbiol 2003; 57:77-100.
47. Wall D, Kaiser D. Type VI pili and cell motility. Mol Microbiol 1999; 32:1-10.
48. Bardy SL, Ng SYM, Jarrell KF. Prokaryotic motility structures. Microbiology 2003; 149:295-304.
49. Manning CD, Schutze H. Foundations of statistical natural language processing. MIT Press, 1999.

Chapter 5

Literature and Genome Data Mining for Prioritizing Disease-Associated Genes

Carolina Perez-Iratxeta,* Peer Bork and Miguel A. Andrade

Abstract

The first step in understanding the molecular biology of an inherited disease is to identify which gene or genes are carrying variants. This process starts with locating the mutations in a chromosomal band, as narrow as possible, and follows with the manual analysis of all the genes mapping in this region. Usually this is not an easy task, but it can be facilitated by complementary computational approaches that evaluate all genes in a region of interest. We present here a method that combines literature mining, gene annotations, and sequence homology searches to prioritize candidate genes involved in a given genetic disorder. The method progresses in two steps. Firstly, we compute associations of molecular and phenotypic features as taken from MEDLINE. Secondly, for a disease with a given phenotype and linked to a chromosomal region, sequence homology based searches are carried on the chromosomal region to identify potential candidates that are scored using the precomputed associations. The scoring of associations between biological concepts using links across databases can be extended to other databases in Molecular Biology and to nondisease phenotypes.

Introduction

Some inherited mutations affecting one or more genes can produce exceptionally grave disorders that affect a high proportion of the population. Well known examples are asthma, diabetes or cancer. Finding out which genes contribute to the phenotypes can open the floor for better therapies, proper diagnosis and prognosis, and even prevention in some cases. Finding genes related to other more rare inherited pathologies has also a high biological and medical importance, because their identification may provide us with new insights about molecular mechanisms, and propitiate medical advances in related areas.

Many genes associated with (mostly monogenic) diseases have been identified and characterized in the past. To date, around 1200 of these are stored in the OMIM database (http://www.ncbi.nlm.nih.gov/omim/). The usual procedure to identify the molecular basis of a monogenic disease is to start by positioning the mutation in the genome by linkage analysis using data from families of affected individuals. The result is a more or less narrow cytogenetic location that is later screened for mutations in genes mapping to the region, often manually selected based on gene function and possible relation to the disease phenotype.

Complex diseases are much harder to position. Weaker linkage correlation signals to loci and the lack of homogeneity within the affected (usually very large) population produce imprecise association to several and larger chromosomal regions. Alternative experimental ways, as the use of polymorphisms, are ongoing (for review see ref. 6).

**Corresponding Author: Carolina Perez-Iratxeta—Ottawa Health Research Institute. Box 411, 501 Smyth Road, Ottawa, Ontario K1H 8L6. Canada. Email: cperez-iratxeta@ohri.ca

Discovering Biomolecular Mechanisms with Computational Biology, edited by Frank Eisenhaber.

Typically, after positional cloning, researchers have to face manual analysis of tens to hundreds of candidate genes, trying to establish possible mechanistical relationships between the ethiology of the disorder and the function of those genes. Even in the case of monogenic diseases, where positional cloning may result in a reasonably narrow band, it can happen that the affected gene has not yet been characterized and functionally annotated, or, in the worst case, has not even been predicted as a gene. Moreover, many diseases are still far from being understood. This means that the molecular biology underlying the particular phenotype is not known, and consequently not described in the biomedical literature.

With the advent of functional genomics approaches, alternative or complementary methodology is frequently used on large sets of genes, for example gene expression analysis using DNA microarrays.[7] Yet, the interpretation of such results is difficult.

We have proposed a computational approach that helps to overcome the three major hampering factors mentioned above, namely, the large size of the cytogenetic bands, the presence of uncharacterized genes therein, and poorly known molecular mechanisms that prevent a straight-forward expert-based selection of candidates.[13]

Our system mines the existing literature and current knowledge about genes, and maps this information to the completed sequence of the human genome. The procedure starts with the phenotype associated to a disease and tries to relate this to molecular functions of the genes in the region. It then scores this information based on a corpus of precomputed links taken from more than 10 Million abstracts in the MEDLINE database. Then it compares the region of interest against annotated proteins by homology search using BLASTX[1a] and produces a ranking according to their scored associations.

Mapping Symptoms to Gene Functions

The first step in our method is to find out automatically which gene functions could be associated to a particular disease phenotype. Both the disease phenotype and the gene function can be summarized with a few keywords describing their main features. Our method is mainly based on detecting the associations between phenotype keywords and gene function keywords.

Our first source of information are the abstracts of literature reports stored in MEDLINE. Each MEDLINE reference is manually annotated with keywords, commonly around a dozen, at the National Library of Medicine (http://www.nlm.nih.gov/). These keywords are organized as an ontology called MeSH (Medical Subjects Headlines, http://www.nlm.nih.gov/mesh/). The MeSH terms are hierarchically organized in eight main categories. The 'C' category, corresponds to 'Diseases'. Then, given a disease, we take as its keywords the MeSH C terms annotated in the MEDLINE references dealing with that disease.

For the genes we use the RefSeq gene database of annotated and validated genes[15] and Gene Ontology[1] as the keyword system. Gene Ontology terms (GO terms, http://www.geneontology.org) constitute an ontology that has become very popular for functional annotation in molecular biology databases. We take as gene keywords all the GO terms associated to a gene in the RefSeq database.

To estimate the degree of relatedness between every MeSH C term, representing a symptom, and every GO term, representing a gene feature, a simple approach could consist of counting how often a given MeSHC term appears in any of the MEDLINE references linked to those gene entries in the RefSeq database annotated with a given GO term.

However, most of the papers linked to RefSeq genes are dealing more with the biochemical characterization of the gene than with clinical matters. Even in the whole MEDLINE there is not enough literature about molecular medicine to permit us to relate symptoms directly to molecular functions. We solve this problem by taking into account that genes relate to phenotypes by means of molecules. Accordingly, we enhance the signal strength of the relations by using an intermediate association step through another MeSH category: the D category consisting of 'Chemicals & Drugs' (see Fig. 1). Then, firstly we count all cooccurrences of MeSH C and MeSH D terms in references of the whole MEDLINE. For example, the MeSH terms 'Brain Ischemia' (C) and 'Glutamic Acid' (D) are mentioned together frequently in MEDLINE

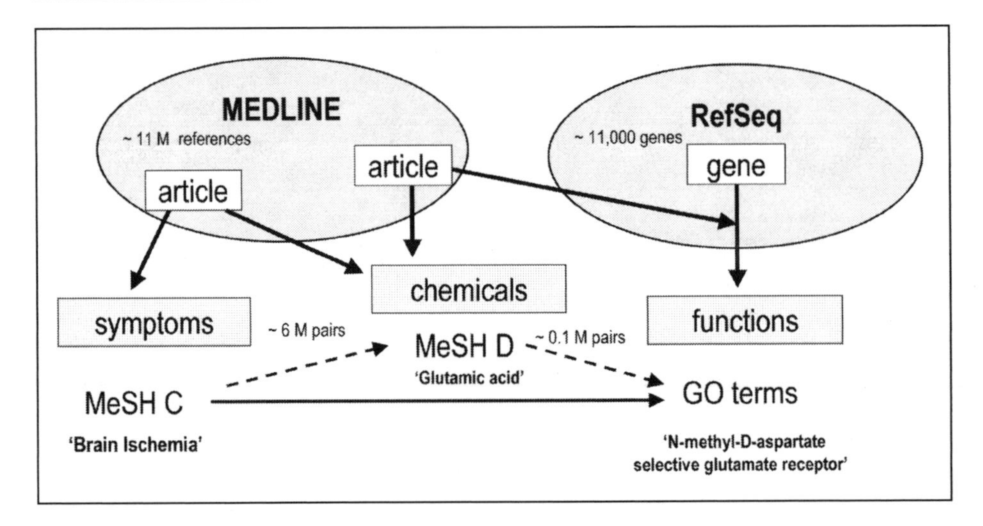

Figure 1. Mapping symptoms to gene features through chemicals.

references. That means that the composed (C,D) pair ('Brain Ischemia','Glutamic Acid') will receive a high value of association. Secondly, we compute the associations between MeSH D terms and GO terms, considering all the bibliography linked in RefSeq. Most of those references will be dealing with the biochemical characterization of the gene. For example, in this way we could find, that 'Glutamic Acid' is strongly associated to the molecular function GO term 'N-methyl-D-aspartate selective glutamate receptor'. The cause for this relation is that many annotations of genes in RefSeq with the GO term 'N-methyl-D-aspartate selective glutamate receptor' are linked to a MEDLINE reference annotated with the MeSH D term 'Glutamic Acid'.

The result of this procedure are two sets of relations: the first one between symptoms and chemicals, and the second one between chemicals and gene function. By combining both sets of relations, it is possible to find associations between phenotypes and gene functions. For formal details about the computation refer to reference 13. Following the example, we could automatically conclude that 'Brain Ischemia' is related to the 'N-methyl-D-aspartate selective glutamate receptor', and the pair receives a high association value. Moreover, the strength of the association can be enhanced by the relation of the symptoms to other terms. In this example, we find that 'Brain Ischemia' is also pointing to the MeSH D term 'Receptors, N-Methyl-D-Aspartate', which turns out to be, logically, highly associated to the 'N-methyl-D-aspartate selective glutamate receptor' GO term.

Once the mapping between keywords is computed, upon collection of the MEDLINE references dealing with a disease, we can extract the most prevailing MeSH C terms (more frequently found in that set of references), and obtain a scoring for all GO terms, according to the learned relation, to the symptoms. We can use the scores to sort all the genes in Refseq, by their annotation with GO terms (for example, with the average of the scores of their GO terms). The resulting score for a gene (GO-score), is giving an indication of the likelihood of the gene of being associated to the given disease (see ref. 13 for details).

Sequence Homology Based Searches

To link a given a chromosomal region to the genes and their associations, we perform BLASTX searches (Altschul et al, 1997) of the region against all the genes in the RefSeq database (Fig. 2). All hits in the region with an E-value below 10e-10 are registered and sorted according to the GO-score of the RefSeq gene they hit. Note that hits in the genome might

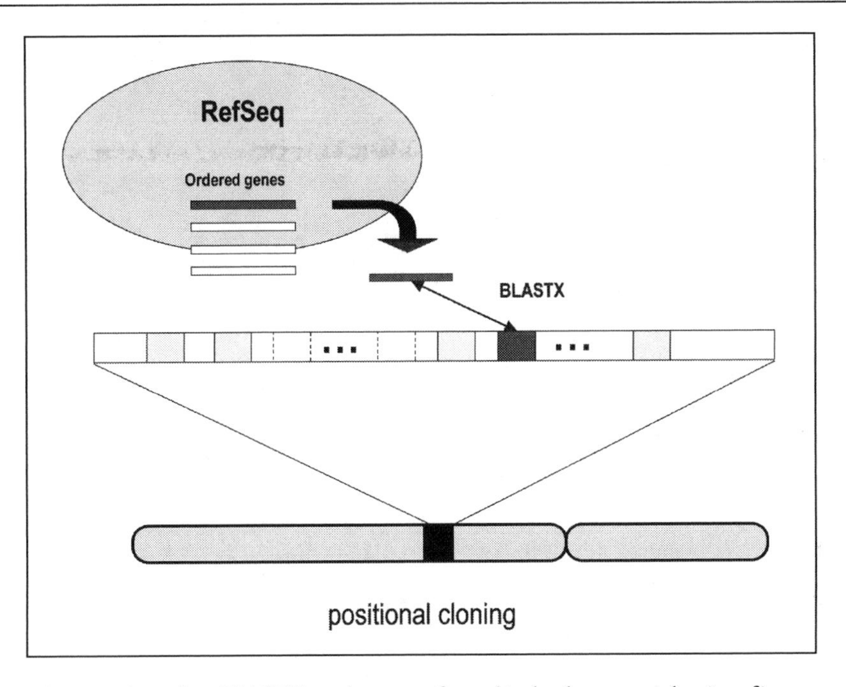

Figure 2. Sequence homology BLASTX searches are performed in the chromosomal region of interest, using the scored genes as queries.

correspond to either known or unknown genes or they might hit a pseudogene.[16] The implications of this fact are discussed below.

If the hits found in the band correspond to genes in RefSeq with the best GO-scores, those tend to correspond with obvious genes that a human expert would point out through manual examination. But it is also possible that none of the genes with high GO-scores gives a hit in the band. In such cases, the first hit can correspond to a gene whose association to the disease is pointed at by a small number of literature references, even a single one. Also, the hit can be obtained by weaker but plausible associations, such as a transcription factor that can be associated with any disease through the gene whose transcription it regulates. To reflect the variable strength of the association of a candidate, we use a relative score (R-score) for the genes in RefSeq that takes into account the distribution of GO-scores. It is computed as the ranking of the gene by GO-score divided by the total number of sequences in RefSeq (1/N for best, and 1 for last, being N the total number of sequences in RefSeq). Homologous genes in the band with an R-score closer to zero indicate "highly anticipated candidates" according to the current knowledge (as extracted from databases by us).

Performance of the System

In order to benchmark the performance of our system, we randomly selected from the LocusLink database[17] 100 genes for which disease-causing mutations had already been reported. In order to obtain the phenotypic terms describing the disease, a set of bibliographic references related to each one was derived, by querying MEDLINE with the name of the disease. However, the associations from these phenotypic terms to chemical terms were computed on a version of MEDLINE from which we removed the 100 sets of references altogether. This was done in order to ensure that we would not find the relation directly from the references that described the connection of a gene to the corresponding disease. A chromosomal region of 30Mb (corresponding to the average size of the regions where unresolved monogenic

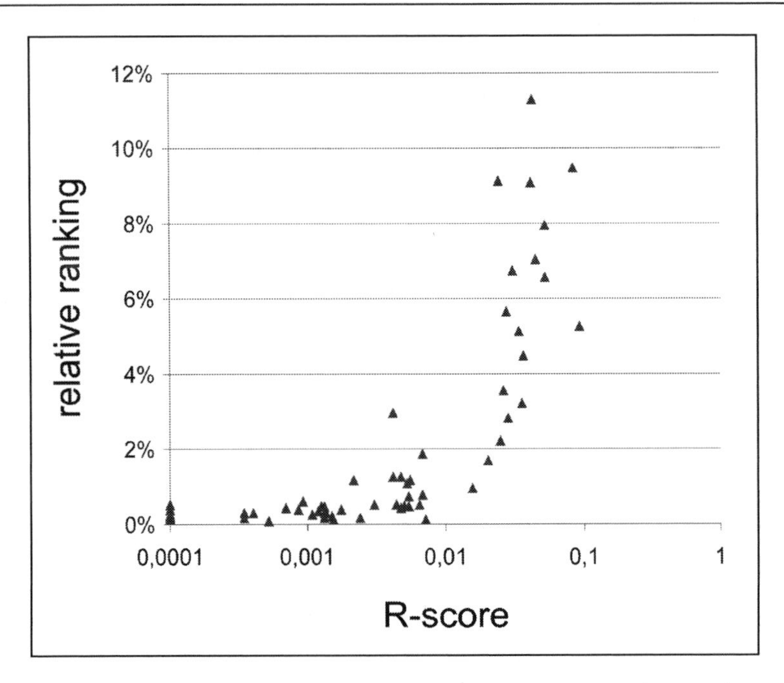

Figure 3. Performance of the system in a benchmark of 100 resolved monogenic diseases. The responsible gene was discovered in 66 cases (represented). The graph displays the relative position of the gene responsible of the disease in the list of candidates versus the R-score of the gene. There is a good correspondence between good (low) R-scores and good rankings. For example, in all but one of the 49 cases where the target gene was scored with an E-score < 0.021, the target gene was detected among the top 2% of the candidates.

diseases were mapped) was taken around each of the 100 genes and the procedure introduced above was applied. All regions with homologues in the RefSeq set that had significant GO-scores were recorded. In an early version of the method,[13] we obtained the following results. The disease-related gene was identified in 55 cases. Of these the disease related gene was among the best 8 scoring genes in 25% of the cases and among the best scoring 30 in 50% of the cases.[13] Recently, we have repeated the same benchmark using updated versions of all the involved databases (manuscript in preparation). Out of the 100 disease-related genes, 61 were identified. In all, the gene was among the 8 best scoring genes in the 43% of the cases, and among the 30 best scoring genes in the last 54% (Fig. 3). We found that the improvement was mainly due to the use of a much more accurate assembly of the human genome (the hg12 , build 30, assembly of the human genome sequence contains nearly 87% finished sequence and 94%-97% coverage) (http://genome.ucsc.edu/).

Examples on How the System Works

We analyzed and made available through the G2D web server (see below) a total of 455 analyses of monogenic diseases. We took some examples of candidates proposed by the system as an illustration of the system's performance.

Often, the prioritization of candidates is straightforward. The obvious criteria are based on the similarity between the prioritized gene and another gene known to produce a slightly different variant of the disease. For example, one of the 455 diseases analyzed was the Charcot-Marie-Tooth disease, axonal, type 2B1 (LocusLink id 65214). It has been linked to chromosome 1q21.2-q21.3.[4] The two equally scoring top candidates are two connexins, the gap junction protein alpha 5 (connexin 40) GJA5 (genbank identifier, gi|6631083) and the gap junction protein alpha 8 (connexin 50) GJA8 (gi|4885275). Both are highly similar to the gap junction protein beta 1 (gi|4504005) which in turn, is known to be responsible of an X-linked Charcot-Marie-Tooth neuropathy.[2] Context analysis reveals that GJA8 is uniquely expressed in eye lense, leaving GJA5 as the most likely candidate.

In other cases, the strong association becomes evident after consulting the literature (yet not easy to retrieve, given the average 300 genes per mapped region). One example is the predicted candidate gene for spinocerebellar ataxia-8 (LocusLink id 3648). The MeSH C most frequently associated to references about this disease is "Spinocerebellar Degenerations". The mapped function terms refer to the neurotransmitter glutamate because glutamate dehydrogenase (GDH), an enzyme central to glutamate metabolism, is significantly reduced in patients with neurological disorders affecting the cerebellum and its connections.[14] Accordingly, the top candidate is human glutamate dehydrogenase 1 located in this region (gi|4885281).

In many cases, however, the criteria for the selection of a candidate gene are less obvious because the potential links in the literature are scarce. One illustrative example was found during our analysis of the dilated cardiomyopathy 1F (LocusLink id 1222), a disease affecting cardiac muscle. The best candidate proposed by the system is a recently cloned cAMP-specific phosphodiesterase 7B (gi|9506959);[8] the relation of the enzyme to the phenotype of the disease is not apparent. The association was originated because inhibitors of phosphodiesterases are used to treat chronic heart failure. Indeed, it has been shown that the decreased gene expression of a cAMP phosphodiesterase is related to dilated cardiomyopathy.[15a]

Sometimes, the associations are weak and may only be based on the similarity of a domain within the protein. Nevertheless given the background of worse associations of other genes in the region, they call for further exploration. An example is the X-linked mental retardation-9 (LocusLink id 4373). Amidst a background of low score hits in the region, the best association to a predicted gene is originated by a strong similarity (and likely orthology) to a murine RNA binding motif protein (gi|6755298). The product of one gene mutated in another X- linked mental retardation disease, the FMR1 gene, is also an RNA binding protein.[10] Although, mechanistic insights into the connection between an RNA- binding function and mental retardation are still missing, the detected candidate gene seems the only reasonable starting point in this region.

A possible example of detection of unknown genes could be the case of a form of epilepsy that is accompanied by mental retardation (LocusLink id 1941). The best-scoring GO terms included references to the neurotransmitter gamma-amino butyric acid (GABA) which, in turn, has been associated to the disease because its levels are affected in several mental retardation diseases (e.g., see ref. 12). Thus, the best-scoring candidate was identified by similarity to GABA receptors. The respective region did not seem to overlap with either a human gene from RefSeq or a predicted gene at the time of the analysis (August 2001). However, since the region matched the entire homologue, did not contain any in-frame stop codon, and was not predicted as a pseudogene, it looked like a plausible candidate. Interestingly, in later assemblies of the human genome sequence, the candidate is overlapping with a gene prediction provided by the Ensembl consortium,[11] in agreement with our previous prediction using homology.

Limitations, Scope, and Further Directions

The major known limitations of our method concern mainly incomplete phenotypical descriptions and the lack or relatively low number of suitable GO annotations for many genes. To overcome the latter, GO terms originating from semi-automatic massive annotation projects such as GOA (Gene Ontology Annotation,[5]) could be used. Another limitation, the sometimes insufficient assembly quality of the genome, seems not to play a big role anymore given the accuracy reached by the current publicly available human genome sequence (http://genome.ucsc.edu/).

Tandem duplications of disease-associated genes are also a factor that hampers a clear assignment as the scoring system introduced here is likely to treat both duplicated genes similarly. Further analysis on the candidates may help, as for example, some Expressed Sequence Tags

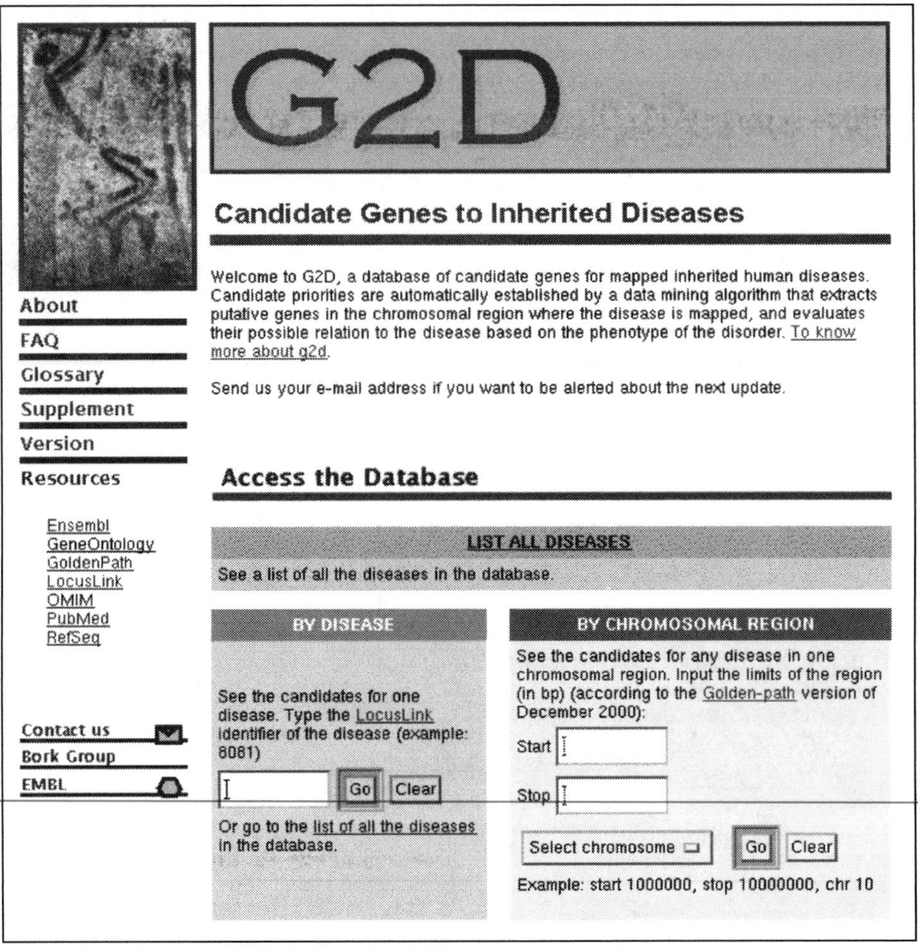

Figure 4. Entry page of the G2D server (http://www.bork.embl.de/g2d2).

(ESTs) or array expression analysis of the duplicated genes might indicate differences in tissue expression helping to identify the correct candidate. Tissue expression based on EST evidence will be provided in the upcoming version of G2D (manuscript in preparation).

It is important to remember that classical pitfalls in function assignment by homology[4] also apply to the procedure presented here. For example, the results of our procedure can occasionally be affected by the "domain problem", that is, a reported high similarity between two sequences is based only on a single shared domain that might cover only a minor portion of both genes. As this can lead to the incorrect assumption that both genes have the same function, the domain organization of the protein has to be considered. Nevertheless, a common domain might be sufficient to capture a true assignment of a candidate gene.

The sequence homology searches may identify unknown and unpredicted genes since it does not rely on the still varying and partially conservative gene prediction pipelines.[9] However, noncoding regions with homology to real genes, such as pseudo-genes, can be wrongly identified as candidates. To discriminate between pseudogenes and undiscovered genes, we are filtering in the upcoming version of G2D for 20,000 predicted human pseudogenes.[16]

Finally, we want to note that in the next version of the system, we offer the users the possibility of scanning any chromosomal region for genes related to disorders present in OMIM. This enables the application of the procedure to complex diseases through the separate analysis of each of several loci. Future work will be devoted to filter the candidate lists using features relating the candidates from multiple loci such as cell pathways, interaction networks, or gene expression data. We plan to integrate other genomic features and new versions of the databases used in future updates of the system. Our goal is to accelerate the path leading geneticists from the genetic linkage of a disease to the characterization of the related gene or genes. Extensive usage of our system and feedback from users will be key to facilitate the evolution of G2D to an optimal tool from the point of view of experimental groups and to take advantage from dynamically evolving resources.

The G2D Web Server

We have developed a publicly accessible web site called G2D (from Genes to Diseases) accessible at the URL http://www.bork.embl.de/g2d/ (Fig. 4). It contains the analysis for up to 500 mendelian inherited diseases that have been linked to a region of the genome but for which the particular associated gene is still unknown. Updated versions of G2D will be placed at the same web site.

References

1. Ashburner M, Ball CA, Blake JA et al. Nat Genet 2000; 25:25-29.
1a. Altschul SF, Madden TL, Schaffer AA et al. Gapped BLAST and PSi-BLAST: A new generation of protein database search programs. Nucleic Acid Res 1997; 25:3389-3402.
2. Bergoffen J, Scherer SS, Wang S et al. Connexin mutations in X-linked charcot-marie-tooth disease. Science 1993; 262:2039-2042.
3. Bork P. Powers and pitfalls in sequence analysis: The 70% hurdle. Genome Res 2001; 10:398-400.
4. Bouhouche A, Benomar A, Birouk N et al. A locus for an axonal form of autosomal recessive Charcot-Marie- Tooth disease maps to chromosome 1q21.2-q21.3. Am J Hum Genet 1999; 65:722-727.
5. Camon E, Magrane M, Barrell D et al. The gene ontology annotation (GOA) database: Sharing knowledge in uniprot with gene ontology. Nucleic Acids Res 2004; 32.
6. Cardon LR, Abecasis GR. Using haplotype blocks to map human complex trait loci. Trends Genet 2003; 19:135-140.
7. Erle DJ, Yang YH. Asthma investigators begin to reap the fruits of genomics. Genome Biol 2003; 4:232.
8. Hetman JM, Soderling SH, Glavas NA et al. Cloning and characterization of PDE7B, a cAMP-specific phosphodiesterase. Proc Natl Acad Sci USA 2000; 97:472-476.
9. Hogenesch JB, Ching KA, Batalov S et al. A comparison of the Celera and Ensembl predicted gene sets reveals little overlap in novel genes. Cell 2001; 106:413-415.
10. Khandjian EW. Biology of the fragile X mental retardation protein, an RNA-binding protein. Biochem Cell Biol 1999; 77:331-342.
11. Lander ES et al. International human genome sequencing consortium. Initial sequencing and analysis of the human genome. Nature 2001; 409:860-921.
12. Olsen RW, Avoli M. GABA and epileptogenesis. Epilepsia 1997; 38:399-407.
13. Perez-Iratxeta, Bork P, Andrade MA. Association of genes to genetically inherited diseases using data mining. Nat Genet 2002; 31:316-319.
14 Plaitakis A, Flessas P, Natsiou AB et al. Glutamate dehydrogenase deficiency in cerebellar degenerations: Clinical, biochemical and molecular genetic aspects. Can J Neurol Sci 1993; 20:109-116.
15. Pruitt KD, Maglott DR. Ref seq and locus link: NCBI gene-centered resources. Nucleic Acids Res 2001; 29:137-140.
15a. Smith CJ, Huang R, Sun D et al. Development of decompensated dilated cardiomyopathy is associated with decreased gene expression and activity of the milrinone-sensitive cAMP phosphodiesterase PDE3A. Circulation 1997; 96:3116-3123.
16. Torrents D, Suyama M, Zdobnov E et al. A genome-wide survey of human pseudogenes. Genome Res 2003; 13:2559-2567.
17. Wheeler DL, Church DM, Edgar R et al. Database resources of the national center for biotechnology information: Update. Nucleic Acids Res 2004; 32:D35-D40.

SECTION III

Mechanistic Predictions from the Analysis of Biomolecular Networks

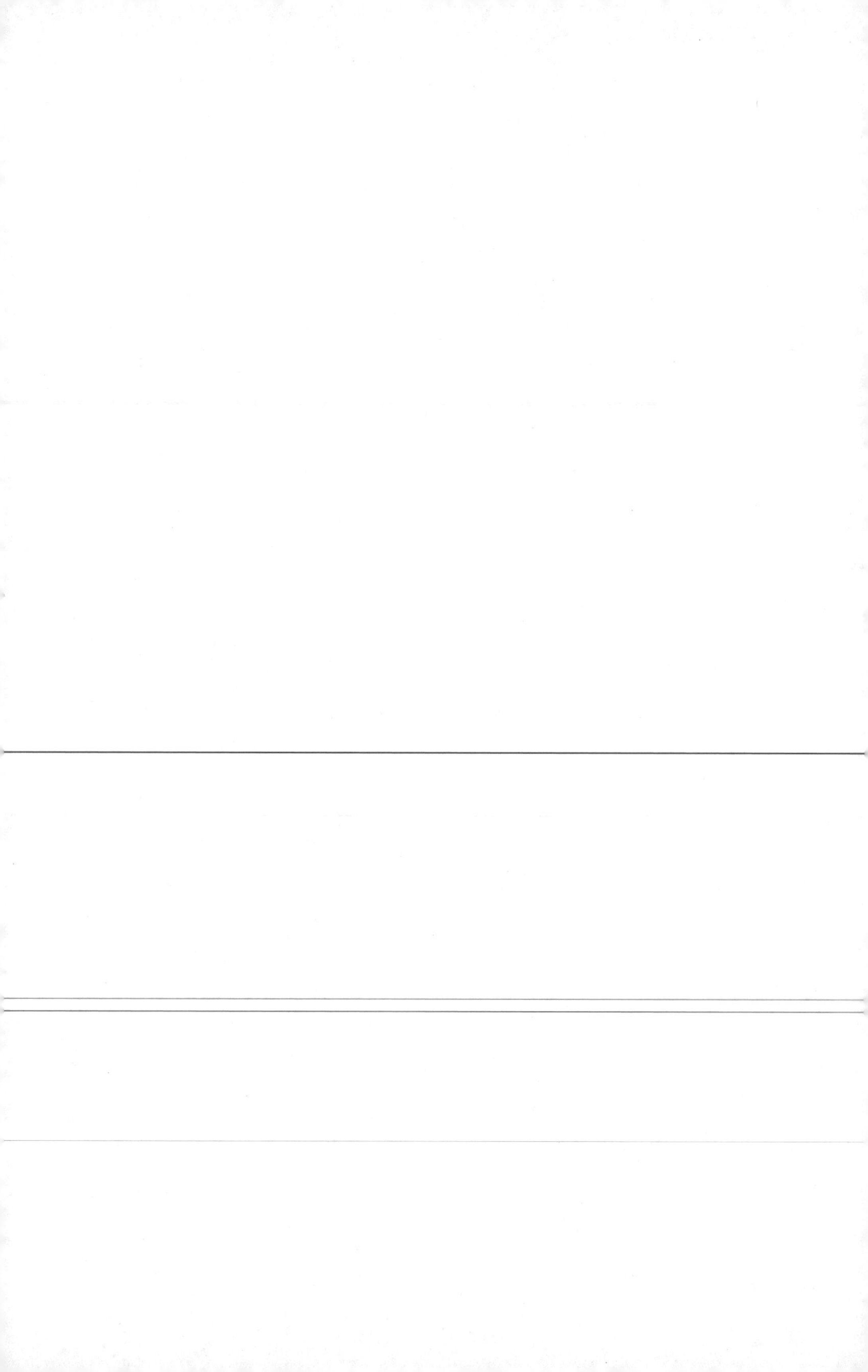

Chapter 6

Model-Based Inference of Transcriptional Regulatory Mechanisms from DNA Microarray Data

Harmen J. Bussemaker*

Abstract

The development of DNA microarray technology has made it possible to monitor the mRNA abundance of all genes simultaneously (the transcriptome) for a variety of cellular conditions. In addition, microarray-based genomewide measurements of promoter occupancy (the occupome) are now available for an increasing number of transcription factors. With this data and the complete genome sequence of many important organisms, it is becoming possible to quantitatively model the molecular computation performed at each promoter, which has as input the nuclear concentration of the active form of various regulatory proteins (the regulome) and as output a transcription rate, which in turn determines mRNA abundance. In this chapter, we describe how our group has used multivariate linear regression methods to: (i) discover cis-regulatory elements in upstream regulatory regions in an unbiased manner; (ii) infer a regulatory activity profile across conditions for each transcription factor; and (iii) determine whether the mRNA expression level of a gene whose promoter is occupied by a particular transcription factor is truly regulated by that factor, through integrated modeling of expression and promoter occupancy data. Together, these results show model-based analysis of functional genomics data to be a versatile conceptual and practical framework for the elucidation of regulatory circuitry, and a powerful alternative to the currently prevalent clustering-based methods.

Introduction

The recent development of high-throughput genomics technologies has had a major impact on the gene expression regulation field. It has become feasible to study the cell from a systems point of view, as a network of interacting genes and their protein products. The genomes of many important model organisms, as well as that of *Homo sapiens*, have been sequenced.[1-4] This has given rise to the development of DNA microarrays as a tool for monitoring the mRNA transcript abundance of all genes in a cell simultaneously,[5,6] and more recently for performing genomewide profiling of the occupancy of noncoding DNA by transcription factors (TFs) using ChIP[7,8] or DamID.[9] The genomewide mRNA expression pattern is commonly referred to as the "transcriptome", while we here propose to refer to the set of genomewide TF occupancies (the terms "binding data" and "location data" are less accurate, in our opinion) as the "occupome".

*Harmen J. Bussemaker—Department of Biological Sciences, and Center for Computational Biology and Bioinformatics, Columbia University, New York, New York, U.S.A.
Email: Harmen.Bussemaker@columbia.edu

Discovering Biomolecular Mechanisms with Computational Biology,
edited by Frank Eisenhaber.

For higher organisms only a small fraction of the genome codes for proteins. The function of the remaining noncoding DNA is largely unknown. It is widely believed that the complexity of an organism crucially depends on the way its genes interact. The unexpectedly low number of protein encoding genes found in the human genome supports this view.[10] However, our understanding of the molecular mechanisms underlying the control of gene expression by regulatory proteins such as transcription factors that bind to noncoding DNA is still very limited, especially for higher eukaryotes. Through integrated analysis of mRNA expression data, transcription factor occupancy data and genome sequence, advances are likely to be made in the mechanistic understanding of the genomewide regulatory network.[11,12]

Two Classes of Tools for Finding Motifs from Expression Data

Knowing the mRNA expression level and the promoter sequence of each gene makes it possible to use computational methods to identify cis-regulatory elements (CREs). Among the various tools used to link gene expression data to cis-regulatory motifs, a fundamental distinction between two classes can be made:

- *Feature enrichment scoring methods ("Class A")* define a subset of genes of interest based on expression data (e.g., all genes upregulated above noise, or the output of a hierarchical clustering algorithm) and subsequently analyze the promoter regions of these genes for overrepresentation of specific sequence patterns.
- *Expression-based feature scoring methods ("Class B")* first define a subset of genes based on a expression-independent feature (e.g., genes whose promoter region contains a specific motif) and subsequently score this feature by comparing the average expression level of the genes containing the feature to the genomewide distribution of expression levels.

Most currently used motif-finding approaches belong to Class A.[13,14] The regression-based methods developed by our laboratory, discussed below, can all be viewed as belonging to Class B. By combining the signals of multiple genes on the microarray, Class B tools have greatly enhanced statistical power to detect differential expression at the level of multi-gene pathways. A change in activity for given transcription factor may be scored as highly significant even if no single gene controlled by that factor shows a change in mRNA expression above noise level.

A similar distinction can be made among methods that aim to combine functional annotation information from Gene Ontology[15] with gene expression data: set enrichment scoring using the cumulative hypergeometric distribution[16] belongs to class A, while methods that score the average expression of genes in each GO category[17,18] belong to class B, and are therefore more sensitive.

REDUCE: Motif-Based Regression Analysis of the Transcriptome

As a first step towards the goal of "reverse engineering" the cell-wide regulatory network from large functional genomics data sets, we recently developed a motif-based regression analysis method named REDUCE, an acronym for "regulatory element detection using correlation with expression".[19] Class A motif finding tools rely on the clustering of genes based on their expression profile across a large number of experimental conditions. By contrast, REDUCE fits a simple model for transcriptional control to a *single* genome-wide expression pattern measured using DNA microarrays. It not only identifies cis-regulatory elements (CREs) in noncoding DNA, but also infers changes in the nuclear concentration of the transcription factors that bind these CREs. Another unique feature of REDUCE is that it naturally takes into account the combinatorial nature of gene expression regulation by allowing multiple factors to control each gene in a unique way, defined by its promoter sequence. Several other groups have adopted and extended our model-based approach.[20-22]

At the core of REDUCE is the following linear model for transcriptional regulation:

$$A_g^{predicted} = \sum_{m \in M} F_m N_{mg}. \tag{1}$$

Here the dependent variable A_g equals the logarithm of the mRNA abundance for gene g, while the independent variable N_{mg} is defined as the number of matches of motif m to the promoter sequence of gene g. For simplicity, A and N are assumed to be normalized to have zero mean. The slopes F_m, estimating the effect that each occurrence of a motif m in the promoter of a gene has on its expression level, are determined by minimizing the deviance D, defined as

$$D = \sum_g \left(A_g^{measured} - A_g^{predicted} \right)^2 \tag{2}$$

The set of motifs M on which the model is based can be determined without any prior knowledge by using forward selection of parameters.[23] Motifs are selected from a large search space (e.g., all oligonucleotides up to 8 base pairs). Each motif is tried as a model parameter, and the motifs are ranked by the value of R^2, the fraction of the total variance explained by the model. The motif with the largest R^2 is then selected as the first motif in the set M, and the procedure is iterated: At each step the residuals based on the current model are calculated, and each motifs in the search space is tested for correlation with the residuals. To avoid over-fitting, a P-value measuring the significance of the increase in R^2 is calculated at each step. The fact that multiple motifs are tested in parallel is accounted for by applying a Bonferroni correction,[24] i.e., by multiplying the single-motif P value by the total number of motifs.

The selection of motifs that correlate with expression may reveal novel motifs or uncover an unexpected role of a known transcription factor in a particular experiment. The latter illustrates the ability of REDUCE to extract condition-dependent information about regulation, which sets it apart from other microarray analysis methods. For instance, REDUCE analysis of a time course for the developmental process of sporulation in *S. cerevisiae*[25] not only confirmed the regulatory role of the known sporulation motifs URS1 and MSE, but also revealed an unexpected role of the mitotic regulator MBF, a heterodimer of Mbp1p and Swi6p binding to the MCB element;[19] the involvement of MBF during meiosis was confirmed by an independently performed genome-wide binding study for MBF and SBF.[8] When used to analyze expression data for a metabolic switch from glucose to oleate in *S. cerevisiae*,[26] REDUCE revealed an unexpected role for the stress-related transcription factors Msn2/4p and Yap1p. The time-dependent subcellular localization of these factors was determined experimentally using a fusion of Msn2p with green fluorescent protein (GFP). The observed transport of this protein, first from the nucleus to the cytoplasm, and later back to the nucleus, was consistent with the transient decrease in regulatory activity predicted for the stress response element (STRE).[26]

These results illustrate the ability of REDUCE to infer condition-specific, protein-level activities of transcription factors, independent of the mRNA-level expression of the TF gene. REDUCE can infer changes in TF activity even if the control is exerted purely post-translationally, e.g., through phosphorylation of the TF. This is a significant advantage over methods that represent activities of transcription factors by mRNA expression levels.[27-29] It is interesting that transcription factor concentrations, which are often as low as 1-10 copies/cell and are therefore hard to measure experimentally, can be more accurately inferred from microarray data using our model-based analysis than the mRNA expression level of individual genes.

From Central Dogma to "Omes Law"

The Central Dogma of biology states that information in the cell flows from DNA to mRNA to protein. Regulatory control is exerted at various levels, including transcription, mRNA turnover and splicing, translation, and protein turnover. The modeling described in this paper is restricted to the very first level of control—i.e., that of transcription initiation—but the regulation of mRNA stability through cis-regulatory elements in the untranslated regions (UTRs) of mRNA has also be addressed within our regression framework (B.C. Foat and H.J. Bussemaker, manuscript in preparation).

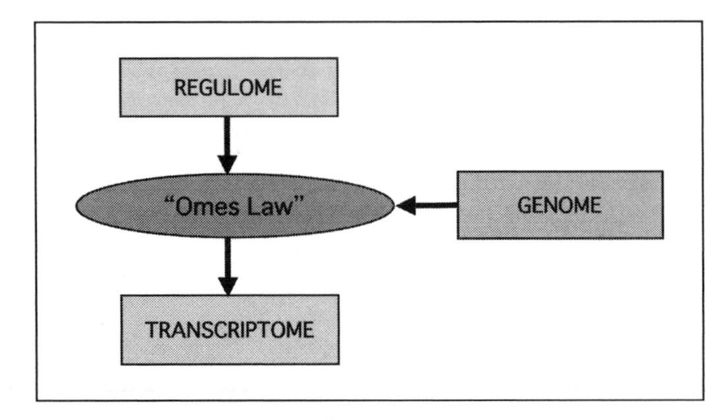

Figure 1. REDUCE as a reverse engineering problem. The transcriptome can be viewed as the output of a complex molecular computation performed at each promoter region (governed by "Omes Law") and dependent on the state of the cell, as represented by the nuclear concentrations of various transcription factors (the "regulome"). When the transcriptome is known experimentally, the unknown regulome parameters can be estimated by fitting the model of Eq. (1).

Equation (1) also ties together information at the level of DNA (as represented by the motif counts N_{mg}), mRNA (as represented by the mRNA expression log-ratios A_g), and protein (as represented by the inferred nuclear transcription factor concentrations F_m). But unlike in the Central Dogma, where the proteins are *downstream* of the mRNA molecules that encode them, in Eq. (1) the proteins are *upstream* of the mRNA transcribed from the genes whose promoters they control.

Figure 1 shows how REDUCE can be viewed as an inverse problem for a process with two inputs and one output: We refer to the regulatory logic of the molecular computation performed by the proteins that occupy each promoter as *Omes Law*. The analogy with the well-known law in electricity does make some sense: the current through a resistor (transcription rate) in response to a voltage difference (transcription factor concentration) depends on the resistance (the strength of binding sites in the promoter region). On a genomewide scale, the transcriptome can be viewed as a consequence of the regulatory state of the cell as represented by transcription factor levels in the nucleus (for which we have previously coined the term "regulome"[19]) and the genome sequence. Analyzing microarray expression data using REDUCE corresponds to a situation where only one of the inputs (i.e., the static genome sequence) and the output (i.e., the dynamic transcriptome) are known, and the other unknown input (i.e., the dynamic regulome) needs be inferred.

Three Different Ways of Using Regression Analysis

As described above, REDUCE uses regression analysis to explain mRNA abundances in terms of motif counts in noncoding regions (black arrow in Fig. 2). The recent emergence of ChIP and DamID for genomewide profiling of transcription factor binding has added a third level of experimental information, and thereby two novel ways of using regression analysis. First, by applying REDUCE to ChIP or DamID log-ratios instead of mRNA expression log-ratios, binding motifs can be found for the profiled transcription factors (bottom arrow in Fig. 2). The value of the regression coefficient, F, depends both on the affinity of the factor for the motif and the concentration of the (DNA-binding form of the) factor in the nucleus in the condition under which the ChIP or DamID experiment was performed. This approach has been used in references 30 and 31. Second, it is possible to use regression analysis to directly relate mRNA expression data to measured TF occupancies (right arrow in Fig. 2). In this case,

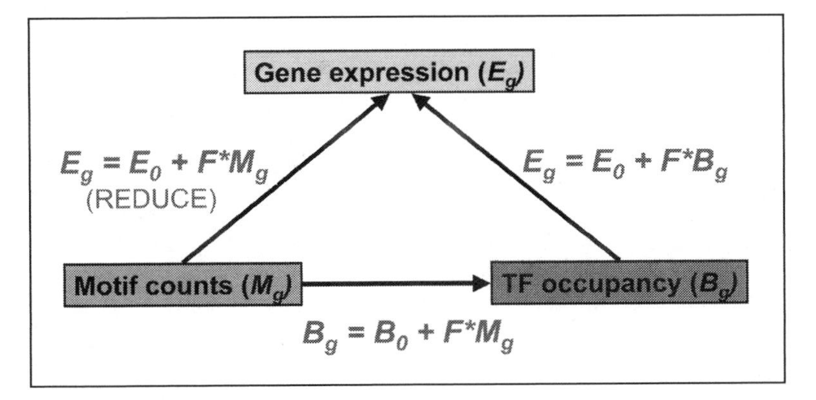

Figure 2. Regression analysis can be used in three different ways to causally link mRNA expression log-ratios, ChIP transcription factor occupancy log-ratios, and motif counts: Cis-regulatory motifs can explain mRNA expression (left arrow) or transcription factor occupancy (bottom arrow); alternatively, transcription factor occupancies can explain mRNA expression directly (right arrow).

the regression coefficient F models the nuclear concentration of the TF under the conditions of the mRNA expression experiment *relative to* the nuclear concentration under the conditions of the ChIP/DamID experiment. This approach forms the basis of the MA-Networker algorithm described below.

MA-Networker: Integrating Occupome and Transcriptome Data

As was shown above, when transcript abundances are compared between experimental conditions in a single microarray experiment, multivariate regression analysis of the mRNA log-ratios can quantify to what extent each transcription factor is responsible for the observed changes in mRNA expression. We have recently generalized the regression approach of REDUCE to infer the genetic regulatory network of the cell from libraries of expression data.[32] We used "ChIP-chip" transcription factor binding data, rather than motif occurrence in promoter regions, as a predictor. If the regression procedure is performed for a large number of conditions ("Step 1" in Fig. 3A), the inferred TF activities can be combined into a highly specific regulatory signature, or transcription factor activity profile (TFAP), for each transcription factor (Fig. 3B).

One expects the mRNA expression profile of a gene that is regulated by a specific transcription factor to be similar to the TFAP of that factor. We therefore investigated whether the linear correlation across the experiment library between a TFAP and the mRNA expression profile of a gene whose promoter is bound by the factor could be interpreted as a regulatory coupling strength and used to improve the specificity of target prediction. To this end, we constructed a matrix of regulatory coupling strengths between all transcription factors and all genes ("Step 2" in Fig. 3A). When this information is combined with the original ChIP data for a given TF, the ChIP log-ratio and coupling strength for each gene can be shown simultaneously in a 2D scatter plot (Fig. 3B).

The biological implications of our results are highlighted in the case of divergently transcribed genes that share a common promoter region, represented as a single microarray probe in the ChIP experiments of reference 33. In this case, our results allow us to identify which genes are controlled by which factors, as shown in Figure 4. Indeed, we found the functional annotation[15] of the protein encoded by the coupled targets to be consistent with what was known about the function of the bound TF in most cases analyzed.

The MA-Networker algorithm outlined in Figure 3 generalizes the linear regression approach of Eq. (1) by taking into account expression data from multiple experimental

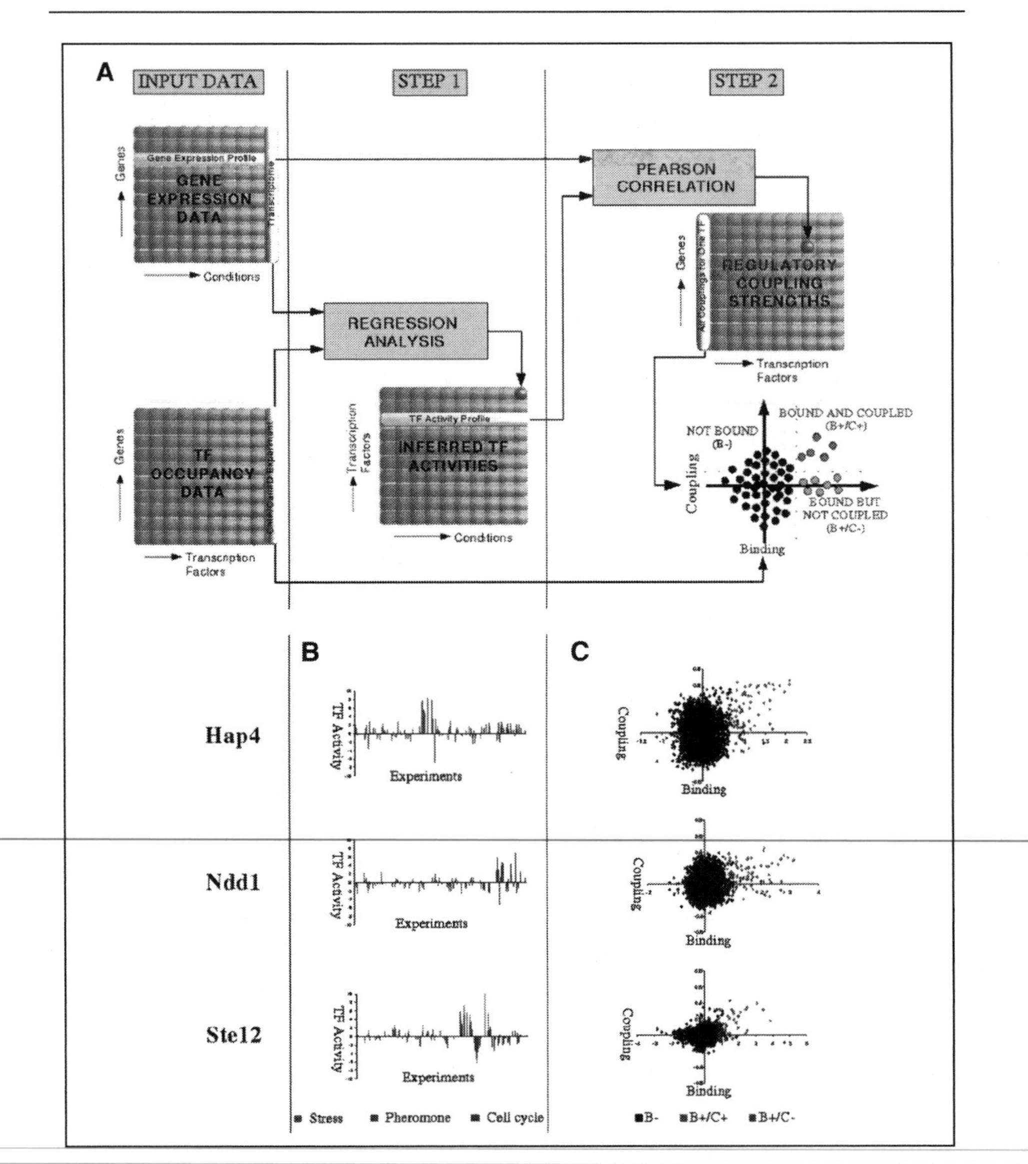

Figure 3. A) Overview of MA-Networker, our method for determining regulatory coupling strengths between transcription factors and their putative target genes. Inputs are (i) a library of microarray expression data for a large number of conditions and (ii) genomewide (ChIP) occupancy data for one or more transcription factors. In the first step of our algorithm, a matrix of transcription factor activities is inferred by using regression analysis to explain the mRNA expression pattern under each condition in terms of the ChIP data for each transcription factor. In the second step, a matrix of regulatory coupling strengths is determined by computing the correlation between each transcription factor activity profile (TFAP) and the mRNA expression profile of each gene. B) Examples of transcription factor activity profiles. Significant changes in activity of the TCA cycle regulator Hap4p occur mostly in metabolic stress conditions, while changes in the activity of the cell cycle regulator Ndd1p and the pheromone-dependent regulator Ste12p are associated with the cell cycle and signal transduction experiments, respectively. C) Scatter plots of ChIP binding log-ratio versus coupling factor. Black dots denote unbound (B-) genes, red dots denote bound and coupled genes (B+/C+), while green dots denote genes that are bound but not coupled (B+/C-). Reproduced from reference 32. A color version of this figure is available online at http://www.Eurekah.com.

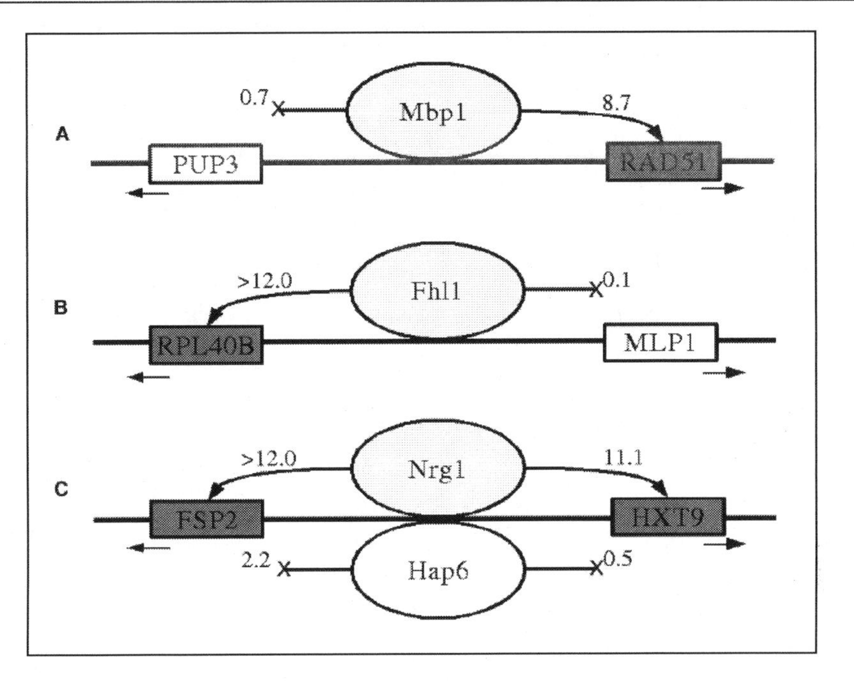

Figure 4. Assigning directionality to divergently transcribed promoters. For pairs of divergently transcribed genes sharing a single promoter region occupied by one or more transcription factors, our method can be used to determine which gene is regulated by which factor. In the diagrams, genes are represented as squares with arrows showing the transcription direction; transcription factors are shown as ovals. The numbers shown are significance scores for the coupling between the transcription factor and the gene, equal to the negative 10-based logarithm of the P-value. Significant regulatory relationships are shown as arrows to colored boxes. Reproduced from reference 32.

conditions simultaneously. To see this, consider the following model for the expression log-ratio A_{gt} for gene g under condition t:

$$A_{gt}^{predicted} = \sum_{f} F_{ft} C_{fg} \tag{3}$$

Here F_{ft} models the activity of transcription factor f under condition t, while C_{fg} represents the extent to which factor f contributes to the regulation of gene g. Note that the matrix F is gene independent. This allows us to determine the TF activities F_{ft} for a given condition t by performing a least-squares fit to the expression data A. A prerequisite for this is that an initial estimate for C_{fg} is available. One possibility is to assume that $C_{fg} = M_{mg}$, where m is a DNA motif to which the factor f is known to bind, and M_{mg} is the number of occurrences of this motif in the promoter region of gene g—this is exactly the assumption that was used in the "standard" version of REDUCE[19] (left arrow in Fig. 2). Another possibility is to assume that $C_{fg} = B_{fg}$, where B_{fg} stands for the ChIP/DamID log-ratio for gene g in an experiment in which the genomewide binding of factor f is measured (bottom arrow in Fig. 2).

Unlike F, the matrix C is condition-independent. Therefore, once we have estimated each element in the matrix F_{ft} by performing a least-squares fit for each individual condition, we can improve our initial guess for the values of C_{fg} by performing a least-squares fit for each gene in which C_{fg} is now treated as a regression coefficient, while F is treated as constant. When this is done for the couplings between a specific transcription factor f and all genes, C_{fg} is proportional to the Pearson correlation between the transcription activity profile (TFAP) for f—i.e., the f-th row from matrix F—and the mRNA profile of gene g—i.e., the g-th row of matrix E_{gt}. In other

Table 1. Glossary

DNA Microarray	Two-dimensional array of spots consisting of immobilized single-stranded DNA probes, which hybridize to fluorescently labeled RNA extracted from a cell culture; the intensity of each spot represents the mRNA abundance for a single gene.
Transcription Factor (TF)	DNA-binding protein that helps recruit the RNA polymerase complex to the start of a gene, and thereby controls the rate at which the gene is transcribed.
ChIP	Chromatin Immuno-Precipitation; experimental technique that produces a mixture of genomic fragments enriched for DNA sequences bound by a specific TF by using antibodies; when used in combination with microarrays ("ChIP-chip"), it can be used to measure protein-DNA interactions on a genomewide scale.
DamID	DNA adenine methyltransferase IDentification; experimental technique that produces data very similar to ChIP, but instead of antibodies uses a fusion between the TF of interest and a DNA adenine methyltransferase (Dam) domain from *E. coli* to obtain an enriched mixture of genomic fragments.
GO	Gene Ontology; an organism-independent, hierarchical classification of all genes based on function, process, and location within the cell; see http://www.geneontology.org.
CRE	Cis-regulatory element; a short DNA region that acts as a transcription factor binding site and plays a role in controlling the transcription of a nearby gene.
TFAP	Transcription Factor Activity Profile; a profile of the regulatory activity of a transcription factor across different experimental conditions; these "hidden variables" can be inferred from mRNA expression data using our regression methods.
Transcriptome	The combined mRNA abundances of all genes for a specific experimental condition.
Occupome	The combined TF occupancies at all promoter regions for a specific transcription factor, obtained under a specific reference condition (often rich media, mid-log phase).
Regulome	The combined TF activities inferred from the transcriptome for a specific experimental condition.

words, the iterative use of regression analysis to determine F and C, respectively, is equivalent to the method summarized in Figure 3.

Conclusion

Analysis of microarray data based on predictive models, as presented here, represents a powerful new paradigm, in which the quality of fit to the data can be used to guide a hypothesis-driven process of model development. Since regulation of chromatin state, transcription initiation, transcript stability, and the interaction between the various regulatory proteins involved in these processes are all amenable to regression analysis, we expect to be able to develop a unified conceptual framework for inferring the structure of the genomewide regulatory network and predicting how the cell responds to changes in environmental conditions or genetic perturbations. We believe our model-based approach is more naturally suited to decyphering the logic of promoter regions than clustering-based methods[34,35] since it models the molecular computation performed at each promoter directly, rather than relying on the existence of sets of similarly expressed genes.

Acknowledgements

Thanks to Ron Tepper for a critical reading of the manuscript. Grant support from the National Institutes of Health (LM007276 and HG003008) is gratefully acknowledged.

References

1. Cherry JM, Ball C, Weng S et al. Genetic and physical maps of Saccharomyces cerevisiae. Nature 1997; 387(6632 Suppl):67-73.
2. Adams MD, Celniker SE, Holt RA et al. The genome sequence of Drosophila melanogaster. Science 2000; 287(5461):2185-95.
3. Lander ES, Linton LM, Birren B et al. Initial sequencing and analysis of the human genome. Nature 2001; 409(6822):860-921.
4. Waterston RH, Lindblad-Toh K, Birney E et al. Initial sequencing and comparative analysis of the mouse genome. Nature 2002; 420(6915):520-62.
5. Schena M, Shalon D, Davis RW et al. Quantitative monitoring of gene expression patterns with a complementary DNA microarray. Science 1995; 270(5235):467-70.
6. Lockhart DJ, Dong H, Byrne MC et al. Expression monitoring by hybridization to high-density oligonucleotide arrays. Nat Biotechnol 1996; 14(13):1675-80.
7. Ren B, Robert F, Wyrick JJ et al. Genome-wide location and function of DNA binding proteins. Science 2000; 290(5500):2306-9.
8. Iyer VR, Horak CE, Scafe CS et al. Genomic binding sites of the yeast cell-cycle transcription factors SBF and MBF. Nature 2001; 409(6819):533-8.
9. van Steensel B, Delrow J, Henikoff S. Chromatin profiling using targeted DNA adenine methyltransferase. Nat Genet 2001; 27(3):304-8.
10. Claverie JM. Gene number. What if there are only 30,000 human genes? Science 2001; 291(5507):1255-7.
11. Banerjee N, Zhang MQ. Functional genomics as applied to mapping transcription regulatory networks. Curr Opin Microbiol 2002; 5(3):313-7.
12. Fickett JW, Wasserman WW. Discovery and modeling of transcriptional regulatory regions. Curr Opin Biotechnol 2000; 11(1):19-24.
13. van Helden J, Andre B, Collado-Vides J. Extracting regulatory sites from the upstream region of yeast genes by computational analysis of oligonucleotide frequencies. J Mol Biol 1998; 281(5):827-42.
14. Tavazoie S, Hughes JD, Campbell MJ et al. Systematic determination of genetic network architecture. Nat Genet 1999; 22(3):281-5.
15. Ashburner M, Ball CA, Blake JA et al. Gene ontology: Tool for the unification of biology. The Gene Ontology Consortium. Nat Genet 2000; 25(1):25-9.
16. Zeeberg BR, Feng W, Wang G et al. GoMiner: A resource for biological interpretation of genomic and proteomic data. Genome Biol 2003; 4(4):R28.
17. Pavlidis P, Lewis DP, Noble WS. Exploring gene expression data with class scores. Pac Symp Biocomput 2002:474-85.
18. Lascaris R, Bussemaker HJ, Boorsma A et al. Hap4p overexpression in glucose-grown Saccharomyces cerevisiae induces cells to enter a novel metabolic state. Genome Biol 2003; 4(1):R3.
19. Bussemaker HJ, Li H, Siggia ED. Regulatory element detection using correlation with expression. Nat Genet 2001; 27(2):167-71.
20. Keles S, van der Laan M, Eisen MB. Identification of regulatory elements using a feature selection method. Bioinformatics 2002; 18(9):1167-75.
21. Wang W, Cherry JM, Botstein D et al. A systematic approach to reconstructing transcription networks in Saccharomy cescerevisiae. Proc Natl Acad Sci USA 2002; 99(26):16893-8.
22. Conlon EM, Liu XS, Lieb JD et al. Integrating regulatory motif discovery and genome-wide expression analysis. Proc Natl Acad Sci USA 2003; 100(6):3339-44.
23. Jobson JD. Applied multivariate regression analysis Volume 1: Regression and Experimental Design. New York: Springer, 1991.
24. Hochberg Y, Benjamini Y. More powerful procedures for multiple significance testing. Stat Med 1990; 9(7):811-8.
25. Chu S, DeRisi J, Eisen M et al. The transcriptional program of sporulation in budding yeast. Science 1998; 282(5389):699-705.
26. Koerkamp MG, Rep M, Bussemaker HJ et al. Dissection of transient oxidative stress response in saccharomyces cerevisiae by using DNA microarrays. Mol Biol Cell 2002; 13(8):2783-94.
27. Segal E, Shapira M, Regev A et al. Module networks: Identifying regulatory modules and their condition-specific regulators from gene expression data. Nat Genet 2003; 34(2):166-76.

28. Ihmels J, Friedlander G, Bergmann S et al. Revealing modular organization in the yeast transcriptional network. Nat Genet 2002; 31(4):370-7.
29. Zhu Z, Pilpel Y, Church GM. Computational identification of transcription factor binding sites via a transcription-factor-centric clustering (TFCC) algorithm. J Mol Biol 2002; 318(1):71-81.
30. van Steensel B, Delrow J, Bussemaker HJ. Genomewide analysis of Drosophila GAGA factor target genes reveals context-dependent DNA binding. Proc Natl Acad Sci USA 2003; 100(5):2580-5.
31. Orian A, Van Steensel B, Delrow J et al. Genomic binding by the Drosophila Myc, Max, Mad/Mnt transcription factor network. Genes Dev 2003.
32. Gao F, Foat BC, Bussemaker HJ. Defining transcriptional networks through integrative modeling of mRNA expression and transcription factor binding data. BMC Bioinformatics 2004; 5:31.
33. Lee TI, Rinaldi NJ, Robert F et al. Transcriptional regulatory networks in Saccharomyces cerevisiae. Science 2002; 298(5594):799-804.
34. Spellman PT, Sherlock G, Zhang MQ et al. Comprehensive identification of cell cycle-regulated genes of the yeast Saccharomyces cerevisiae by microarray hybridization. Mol Biol Cell 1998; 9(12):3273-97.
35. Beer MA, Tavazoie S. Predicting gene expression from sequence. Cell 2004; 117(2):185-98.

CHAPTER 7

The Predictive Power of Molecular Network Modelling:

Case Studies of Predictions with Subsequent Experimental Verification

Stefan Schuster,* Edda Klipp and Marko Marhl

Abstract

Since the 1960s, the mathematical modelling of intracellular systems, such as metabolic pathways, signal transduction cascades and transport processes, is an ever-increasing field of research. The results of most modelling studies in this field are in good qualitative or even quantitative agreement with experimental results. However, a widely held view among many experimentalists is that modelling and simulation only reproduce what has been known before from experiment. A true justification of theoretical biology would arise if theoreticians could predict something unknown, which would later be found experimentally. Theoretical physics has achieved this justification by making many right predictions, for example, on the existence of positrons. Here, we review three cases where experimental groups that were independent of the theoreticians who had made the predictions confirmed theoretical predictions on features of intracellular biological systems later. The three cases concern the optimal time course of gene expression in metabolic pathways, the operation of a metabolic route involving part of the tricarboxylic acid cycle and glyoxylate shunt, and the decoding of calcium oscillations by calcium-dependent protein kinases.

Introduction

The mathematical modelling and simulation of intracellular biological systems, such as metabolic networks, signalling cascades and transport processes, has become a flourishing field of biological research. This field can be traced back to the work by Henri[1] and Michaelis and Menten[2] on enzyme kinetics, yet the proper start of it should probably be dated in the 1960s, with the work by Garfinkel and Hess,[3] Higgins[4] and others.

Modelling and simulation has manifold purposes, the most important being

- Fitting of experimental data by phenomenological equations
- Fitting of experimental data by equations based on mechanistic knowledge and, thus, explanation of these data
- Planning of experiments
- Replacement of expensive or ethically problematic experiments
- Prediction of hitherto unknown phenomena.

*Corresponding Author: Stefan Schuster—Friedrich Schiller University Jena, Faculty of Biology and Pharmaceutics, Section of Bioinformatics, Ernst-Abbe-Platz 2, D-07743 Jena, Germany. Email: schuster@minet.uni-jena.de

Discovering Biomolecular Mechanisms with Computational Biology, edited by Frank Eisenhaber.

The latter purpose is obviously the most ambitious goal. However, it is achieved relatively rarely in theoretical biology. Although the acceptance of theoretical (in particular, mathematical) biology appears to increase, as witnessed by the increasing number of such papers in high-ranking journals, experimentalists are often somewhat hesitant with respect to modelling. Many of them believe that it only reproduces what has been known before from experiment. On the other hand, all biochemists make use of the Michaelis-Menten kinetics, which is a form of modelling.

By contrast, theoretical physics is a much more established discipline because it has allowed the prediction of many phenomena that have been found later by experiment. For example, the positron was predicted by P.A.M. Dirac in 1931 and found by C.D. Anderson in 1932 (cf. ref. 5). More recently, powerful methods of theoretical physics, in particular using analogies, enabled a considerable progress in the field of liquid crystals. The analogy between superconductors and the liquid crystal smectic-A phase, found by Nobel Prize winner P.G. de Gennes,[6] turned out to be an extremely powerful tool. By knowing the properties of the normal metal—superconductor phase transition, several properties of the nematic—smectic-A phase transition could be foreseen. The use of the analogy led to the prediction and theoretical description of the twist grain boundary phases[7] (TGB phases). This phase is a liquid crystal analog to the Abrikosov (Nobel Prize winner in 2003) lattice in superconductors.[8] One year after the TGB phase was predicted, an unusual phase was observed in smectic-A liquid crystals.[9] It turned out that the theoretically predicted TGB phase had been discovered experimentally.

In theoretical biology, the successful prediction of phenomena unknown earlier is much less frequent than in physics. What comes to mind is the prediction of three-dimensional structures of proteins, which is often successful, but often it is not.[10] However, this prediction only assigned well-known structure elements such as α-helix and β-sheet to proteins for which the structure had not been known before rather than yielding completely new structures.

Here, we review three cases where theoretical predictions of phenomena or features in intracellular biological systems have indeed been verified later by experimentalists that worked independently of the theoreticians who made the predictions.

The first reported case concerns the prediction of properties of metabolic systems from optimality principles. Biological systems developed through evolution by mutation and selection. Evolution is often considered as an optimisation process that took place over millions of years. However, evolution is almost always coevolution, that is, different species interact so that they cannot optimise their properties in isolation. Therefore, to explain the unlimited possible forms and strategies, approaches more complex than simple optimisation, such as evolutionary game theory, should be used.[11,12] Evolutionary game theory has, for example, been applied to biochemical systems.[13] Fortunately, there are cases where the game-theoretical problem can be transformed into an optimisation problem.[12] Indeed, it seems that many present-day intracellular systems show properties that are optimal with respect to certain selective conditions.[14,15] Hence, system properties may be predicted from mathematical models based on optimality criteria. Several optimality criteria have been proposed. For cellular reaction systems they involve, for example: (i) maximisation of steady-state fluxes,[14,16-18] (ii) minimisation of the concentrations of metabolic intermediates,[14] (iii) minimisation of transition times,[14] (iv) maximisation of sensitivity to an external signal,[19] (v) optimisation of thermodynamic efficiencies,[20] and (vi) minimisation of total enzyme concentration.[21] Here, we discuss the prediction of temporal expression profiles for genes coding for the enzymes of a pathway from the criterion of minimal time required for the conversion of the pathway's substrate into its product. As a side constraint, the limited capacity of the cell to produce and store proteins is taken into account.

The second case is in the field of Metabolic Pathway Analysis. The set of linear pathways in biochemistry textbooks often does not capture the full range of possible behaviours of a metabolic network. A well-known pathway is the tricarboxylic acid (TCA) cycle. It has frequently been realized that in many organisms only part of this cycle is operative, always or under specific conditions.[15,22] The question arose how the full set of potential pathways in the system could be determined in a systematic way. Various methods have been developed to answer this question (for reviews, cf. refs. 23 and 24). By these methods, a plethora of pathways have been predicted, some of which had been unknown before or had attracted little interest. Here, we report the prediction of one specific pathway involving part of the TCA cycle that has later been verified in experiment.

The third case is in the field of calcium oscillations. In living cells, upon cell stimulation by an agonist, like a hormone or neurotransmitter, often oscillatory changes in free cytosolic calcium concentration are evoked. The so-called Ca^{2+} oscillations play an important role in intra- and intercellular signalling. Many cellular processes like cell division, cell secretion, and egg fertilisation are regulated by Ca^{2+} oscillations. Soon after the discovery of such oscillations in nonexcitable cells[25] (in excitable cells such as neurons they had been known for a longer time), it has been shown that the response of a cell stimulated by different concentrations of a hormone is characterised by different frequencies of Ca^{2+} oscillations.[26] The idea of frequency-encoded Ca^{2+} signals was born and the mechanism of information encoding in the frequency of Ca^{2+} oscillations has been studied theoretically, starting already by the first model of Ca^{2+} oscillations in nonexcitable cells[27] (for reviews of later models see refs. 28,29).

The result caused by the oscillatory Ca^{2+} signal is, in most cases, a (nearly) stationary output, for example, the enhancement of the expression of a gene. For a long time, it had been unclear how this transformation is brought about. Here, we discuss a mathematical model for explaining this and the subsequent experimental proof.

Optimal Time Course of Gene Expression

One of the authors studied, by mathematical modelling, the time course of the adaptation of enzyme concentrations for a pathway that can be regulated depending on the actual requirements.[30] The product, P, of the pathway is considered to be important, but not essential for the reproduction of the cell. The calculation is based on the widely used assumption that during evolution, pathway flux has been maximized.[14,16,18] The faster the initial substrate, S_0, can be converted into P, the more efficiently the cell may reproduce and out-compete other individuals. If S_0 is available, then the cell produces the enzymes of the pathway to make use of the substrate. If the substrate is not available, then the cell does not synthesize the respective enzymes for economical reasons. This scenario is studied theoretically by starting with a resting pathway, i.e., although the genes for the enzymes are present, they are not expressed due to lack of the substrate. Suddenly S_0 appears in the environment (by feeding or change of place). How can the cell make use of S_0 as soon as possible?

For simplicity's sake, linear rate laws are used. The system of differential equations describing the dynamics of the pathway then reads

$$\frac{dS_0}{dt} = -k_1 \cdot E_1 \cdot S_0 \tag{1a}$$

$$\frac{dS_i}{dt} = k_i \cdot E_i \cdot S_{i-1} - k_{i+1} \cdot E_{i+1} \cdot S_i \qquad (i = 1,\ldots,n-1) \tag{1b}$$

$$\frac{dP}{dt} = k_n \cdot E_n \cdot S_{n-1} \tag{1c}$$

Moreover, it is assumed that the cell can synthesize the enzymes instantaneously when necessary (neglecting the time necessary for transcription and translation), but that the total amount of enzyme is limited due to limited capacity of the cell to produce and store proteins. The time necessary to produce P from S_0 is measured by the transition time

$$\tau = \frac{1}{S_0(0)} \int_{t=0}^{\infty} \left(S_0(0) - P(t)\right) dt. \tag{2}$$

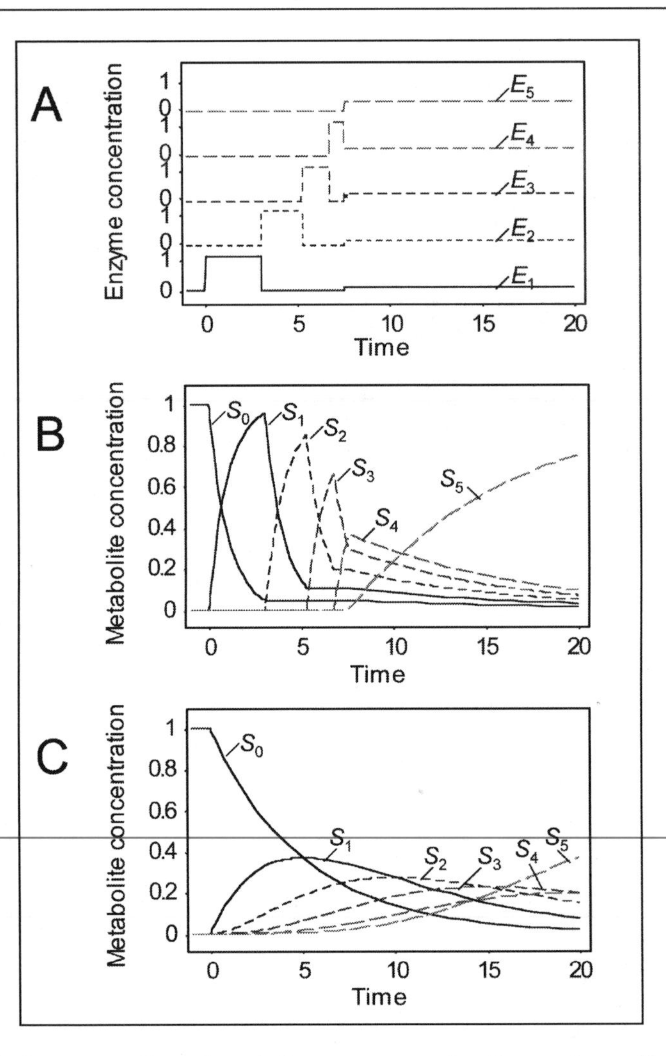

Figure 1. Optimal enzyme profiles and metabolite concentrations for a linear pathway of five reactions. A) Enzymes are switched on and off consecutively (E_i, enzyme concentrations, with the numbering starting at the enzyme converting the initial substrate, S_0). After the last switch, all enzymes get a share of the total enzyme amount. B) The respective optimal time courses of metabolite concentrations, S_i. Comparison to panel A shows that the enzymes are always present, when there is metabolite to degrade. C) Time courses of metabolite concentrations for the reference case that all enzymes get the same share of the total enzyme amount, $E_i(t) = E_{tot}/5$. Comparison to panel B shows that the production of the final metabolite S_5=P is much slower than in the optimal case.

The optimisation problem to be solved is to find a temporal profile of enzyme concentrations that allows for

$$\tau = \min \text{ subject to } E_{tot} = \sum_{i=1}^{n} E_i(t) = const. \tag{3}$$

Varner and Ramkrishna[31] had proposed a similar optimisation criterion, though mathematically less elaborate. By standard optimisation techniques, optimal enzyme profiles can be derived.[30] Figure 1 shows the optimal enzyme profiles and the time courses of metabolite concentrations for the optimal case and for the case of even distribution of enzyme concentrations among reactions. The optimal enzyme profiles have the following characteristics: within successive time intervals, only a single enzyme is fully active whereas all others are shut off. At the beginning of the process, the whole amount of available protein is spent exclusively on the first enzyme of the chain. Each of the following switches turns off the active enzyme and allocates the total amount of protein to the enzyme that catalyses the following reaction. The last switch allocates a finite fraction of protein to all enzymes with increasing amounts from the first one to the last one. This scenario ensures that the enzymes are present when their substrates are present.

Bacterial amino acid production pathways seem to be regulated in the same manner: Zaslaver and colleagues investigated experimentally the amino-acid biosynthesis systems of *Escherichia coli*, identified the temporal expression pattern and showed a hierarchy of expression that matches the enzyme order predicted for unbranched pathways.[32]

Well-characterized enzymatic pathways carry out amino-acid biosynthesis in *E. coli*. To study design principles of metabolic regulation, the promoter activity of about 100 genes have been measured in parallel using GFP and Lux reporter libraries. The dynamics of pathway activation has been analysed by diluting the cells in a defined medium that contains all amino acids except one (e.g., arginine, serine, or methionine). A temporal order of expression has been found with delays of the order of 10 min between different promoters. This temporal order matches the functional enzyme order in the different unbranched pathways of amino-acid biosynthesis.

In addition, it has been found that the closer an enzyme is to the upper end of the pathway, the higher its maximal promoter activity will be. This is in accordance with the theoretical prediction[17,21] that in optimal time-independent states (maximal flux at fixed total enzyme concentration or, in this case equivalently, minimal total enzyme concentration ensuring a fixed flux) the total amount of enzyme along a pathway is distributed to the individual enzymes according to the flux control of these enzymes. Under the simplifying assumption that all enzymes would have similar kinetic properties, enzymes located at the beginning of the pathway have more flux control then those at more downstream positions[33] and should, hence, have a higher concentration.

A Previously Unknown Metabolic Pathway

A basic concept in Metabolic Pathway Analysis is that of elementary flux modes.[34-36] Elementary flux modes are minimal sets of enzymes that can operate together at steady state such that the irreversible enzymes involved are used in the right direction. Schilling and coworkers[37] proposed the related concept of extreme pathways. Elementary modes can be described by flux vectors, V, fulfilling three properties:

i. They allow the system to be at steady state. This means, in mathematical terms, that the product of the stoichiometry matrix and the flux vector equals the null vector, $N\,V = 0$.
ii. The components of the flux vector corresponding to irreversible reactions are nonnegative.
iii. There is no flux vector satisfying conditions (i) and (ii) corresponding to a proper subset of the enzymes that correspond to the flux vector V in question.

One of the authors was involved in a study of the pathway structure in the central metabolism of *E. coli*.[35] We analysed a reaction system involving 24 reactions. They correspond to the TCA cycle, glyoxylate shunt and some adjacent reactions of amino acid synthesis. The system gives rise to 26 elementary modes. One of these (mode 6 according to the numbering in ref. 35) involves the enzymes shown in Figure 2. This mode is a combination of the glyoxylate shunt (isocitrate lyase and malate synthase) with part of the TCA cycle and involves, in addition, phosphoenolpyruvate (PEP) carboxykinase. The oxaloacetate (OAA) produced by malate dehydrogenase is used in equal proportions by PEP carboxykinase and citrate synthase. The overall reaction is

$$\text{ADP} + \text{FAD} + 4\ \text{NAD} + \text{2-phosphoglycerate} \rightarrow \text{ATP} + \text{FADH}_2 + 4\ \text{NADH} + 3\ \text{CO}_2.$$

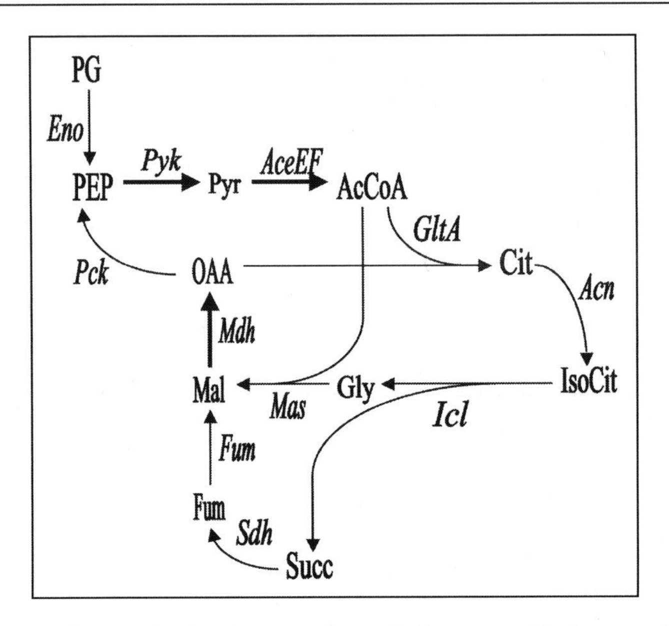

Figure 2. Pathway involving enolase (Eno), pyruvate kinase (Pyk), pyruvate dehydrogenase (AceEF), citrate synthase (GltA), aconitase (Acn), isocitrate lyase (Icl), malate synthase (Mas), succinate dehydrogenase (Sdh), fumarase (Fum), malate dehydrogenase (Mdh), and phosphoenolpyruvate carboxykinase (Pck). Thick arrows correspond to fluxes that are double as high as the other fluxes. 2-Phosphoglycerate (PG) is considered as external metabolite (a source in this case). Cofactors and CO_2 are not displayed.

Fischer and Sauer[38] have found experimentally the above-mentioned pathway in *E. coli* under conditions of glucose hunger (suboptimal supply of the nutrient, glucose). Note that carbohydrates can be oxidized completely to CO_2 and water by this pathway. Thus, in contrast to earlier views, PEP carboxykinase is not only used for anaplerosis, nor is the glyoxylate shunt used for gluconeogenesis only. As the enzymes involved in the new metabolic cycle are present in many microorganisms, it can be speculated that it operates also in other microbes. Its ATP yield per mole of phosphoglycerate is smaller than that of the usual TCA cycle (1 instead of 2). The advantage appears to be the higher NADH yield (4 rather than 3).

The pathway in question is part of a larger pathway predicted for *E. coli* earlier by Liao and coworkers.[39] They studied the synthesis of DAHP, a precursor of aromatic amino acids. The larger pathway involves, in addition, the pentose phosphate pathway and DAHP synthase. Liao and coworkers[39] observed that the enzymes of the glyoxylate shunt and PEP carboxykinase are not highly expressed under glucose-rich conditions and speculated that their levels in nongrowing cells may be higher. This was confirmed by Fischer and Sauer.[38]

Decoding of Calcium Oscillations

At the end of the 1980s, researchers in the field of Ca^{2+} oscillations thought about the question how frequency encoded signals can be decoded into a frequency-dependent cellular response by the targets in nonexcitable cells. This question was first tackled by Goldbeter and coworkers,[40] inspired by earlier ideas concerning activation of the calcium/calmodulin-dependent protein kinase type II by repeated Ca^{2+} peaks in nerve cells.[41,42] This kinase (CaM kinase II, EC 2.7.1.123) is a ubiquitous, multifunctional enzyme. In the pioneering studies on decoding of Ca^{2+} oscillations, Goldbeter and coworkers[40,43] suggested protein phosphorylation to be a possible mechanism for frequency decoding of Ca^{2+} oscillations.

The general model for frequency decoding of Ca^{2+} oscillations[40] takes into account a phosphorylation-dephosphorylation cycle with a Ca^{2+}-activated kinase and a (normally

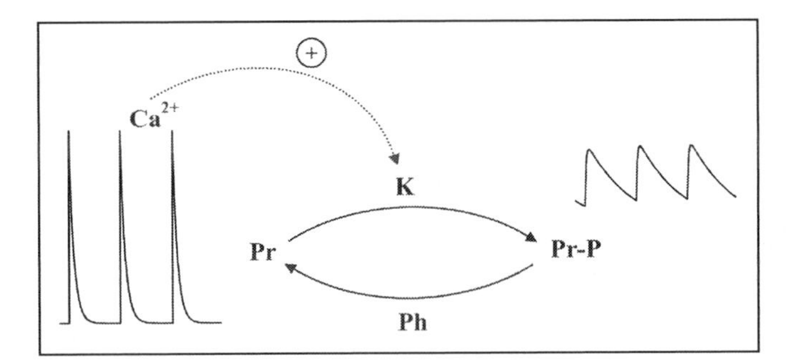

Figure 3. Protein phosphorylation regulated by Ca^{2+} oscillations. Protein (Pr) is phosphorylated by a Ca^{2+}-activated kinase (K). The phosphorylated protein (Pr-P) is dephosphorylated by a phosphatase (Ph).

Ca^{2+}-independent) phosphatase (Fig. 3). The role of CaMKII in decoding Ca^{2+} oscillations has been studied by theoretical models, which have predicted that the level of activity of CaMKII (or of other suitable proteins) increases with the frequency of Ca^{2+} oscillations due to a sort of time integration of the oscillatory signal.[43-47]

Although those theoretical studies have predicted that CaMKII could act as a frequency decoder of Ca^{2+} oscillations, experimental evidence was lacking. De Koninck and Schulman[48] were the first who tested this prediction experimentally, demonstrating that CaMKII is indeed able to decode the frequency of Ca^{2+} oscillations. The importance of this finding and the exact relation between the experimental study and the theoretical models have been discussed (in refs. 49 and 50).

Whereas the basic theoretical studies consider a phosphorylation-dephosphorylation cycle with a Ca^{2+}-dependent kinase and a phosphatase (see above), the results by DeKoninck and Schulman[48] show that the isolated CaMKII itself can act as a frequency decoder, at least over a limited period of time. Nevertheless, phosphatases are important in determining the asymptotic sensitivity of a Ca^{2+} decoding system because of the mass balance of phosphorylated protein.[50,51] Moreover, it should be noted that the experimental study by DeKoninck and Schulman[48] is an in vitro study, which does not take into account a counteracting role played in vivo by the phosphatases. The effect of phosphatases has been studied theoretically by Hanson and coworkers[45] and, again, their results should stimulate further experimental studies to verify the model predictions about the frequency decoding of Ca^{2+} oscillations in vivo.

Discussion

Here, we have discussed three cases of theoretical predictions in cell biology which were later proved by experiment. These cases concern different phenomena in cell physiology: gene expression, metabolism, and signalling. Our case studies illustrate the importance of mathematical modelling. None of the predictions discussed here would have been possible without it. For example, although the concept of biochemical pathway has been used by biochemists for a long time on an intuitive basis, the prediction of stoichiometrically and thermodynamically feasible routes in metabolism only became possible when mathematical methods had been introduced in the field.

Of course, many theoretical predictions have been made that still await their experimental confirmation. One example is the role of phosphatases mentioned in the previous section. Moreover, experimental studies are needed to verify predictions about the bistability behaviour of the CaMKII. The switch between a low activity state and a high activity state, once the Ca^{2+} stimulation exceeds a threshold, could be physiologically very important, in particular, in the brain for the long-term storage of information.[42,52] The occurrence of an all-or-none switch between ON and OFF states of CaMKII activity is predicted to depend on the frequency of Ca^{2+} oscillations.[49]

The first case discussed here—the optimal regime of gene expression—can be considered as a proof of concept of the optimality principle in biology. This principle has sometimes been criticized by saying it can finally be reduced to a tautology. Indeed, to any observed, quantifiable phenomenon, one can define, *a posteriori*, a mathematical function that is maximized by this phenomenon. If, however, verifiable theoretical predictions can be derived from optimality principles, reasoning in terms of optimality must be more than a tautology.

Acknowledgements

The authors would like to thank Dr. Geneviève Dupont (Brussels) for helpful comments on the section *Decoding of Calcium Oscillations* and Drs. Reinhard Meinel (Jena) and Nataša Vaupotič (Maribor) for stimulating discussions on theoretical predictions in physics. Moreover, financial support from the German Ministry of Research and Education to E.K. via the Berlin Center for Genome based Bioinformatics and to S.S. and M.M. by travel grants for mutual visits is gratefully acknowledged.

References

1. Henri MV. Théorie générale de l'action de quelques diastases. Compt Rend Acad Sci 1902; 135:916-919.
2. Michaelis L, Menten ML. Die Kinetik der Invertinwirkung. Biochem Z 1913; 49:333-369.
3. Garfinkel D, Hess B. Metabolic control mechanisms. VII. A detailed computer model of the glycolytic pathway in ascites cells. J Biol Chem 1964; 239:971-983.
4. Higgins J. A chemical mechanism for oscillation of glycolytic intermediates in yeast cells. Proc Natl Acad Sci USA 1964; 51:989-994.
5. Weinberg S. The Quantum Theory of Fields. Vol I. Cambridge: Cambridge University Press, 1995.
6. De Gennes PG. Analogy between superconductors and smectics-A. Solid State Comm 1972; 10:753-756.
7. Renn SR, Lubensky TC. Abrikosov dislocation lattice in a model of the cholesteric to smectic-A transition. Phys Rev A 1988; 38:2132-2147.
8. Tinkham M. Introduction to Superconductivity. New York: McGraw-Hill, 1975.
9. Goodby JW, Waugh MA, Stein SM et al. Characterization of a new helical smectic liquid-crystal. Nature 1989; 337:449-452.
10. Eisenhaber F, Persson B, Argos P. Protein structure prediction: Recognition of primary, secondary, and tertiary structural features from amino acid sequence. Crit Rev Biochem Mol Biol 1995; 30:1-94.
11. Hofbauer J, Sigmund K. Evolutionary Games and Population Dynamics. Cambridge: Cambridge University Press, 1998.
12. Levin SA, Muller-Landau HC. The emergence of diversity in plant communities. CR Acad Sci Paris Life Sci 2000; 323:129-139.
13. Pfeiffer T, Schuster S. Game-theoretical approaches to studying the evolution of biochemical systems. Trends Biochem Sci 2005; 30:20-25.
14. Heinrich R, Schuster S. The Regulation of Cellular Systems. New York: Chapman & Hall, 1996.
15. Meléndez-Hevia E, Waddell TG, Cascante M. The puzzle of the Krebs citric acid cycle: Assembling the pieces of chemically feasible reactions, and opportunism in the design of metabolic pathways during evolution. J Mol Evol 1996; 43:293-303.
16. Kacser H, Beeby R. Evolution of catalytic proteins. On the origin of enzyme species by means of natural selection. J Mol Evol 1984; 20:38-51.
17. Heinrich R, Klipp E. Control analysis of unbranched enzymatic chains in states of maximal activity. J theor Biol 1996; 182:243-252.
18. Stephani A, Nuño JC, Heinrich R. Optimal stoichiometric designs of ATP-producing systems as determined by an evolutionary algorithm. J theor Biol 1999; 199:45-61.
19. Goldbeter A, Koshland Jr DE. Ultrasensitivity in biochemical systems controlled by covalent modification. Interplay between zero-order and multistep effects. J Biol Chem 1984; 259:14441-14447.
20. Stucki JW. The optimal efficiency and the economic degrees of coupling of oxidative phosphorylation. Eur J Biochem 1980; 109:269-283.
21. Brown GC. Total cell protein concentration as an evolutionary constraint on the metabolic control distribution in cells. J theor Biol 1991; 153:195-203.
22. Huynen MA, Dandekar T, Bork P. Variation and evolution of the citric-acid cycle: A genomic perspective. Trends Microbiol 1999; 7:281-291.
23. Papin JA, Price ND, Wiback SJ et al. Metabolic pathways in the post-genome era. Trends Biochem Sci 2003; 28:250-258.

24. Schuster S. Metabolic pathway analysis in biotechnology. In: Kholodenko BN, Westerhoff HV, eds. Metabolic Engineering in the Post Genomic Era. Wymondham: Horizon Scientific, 2004:181-208.
25. Cuthbertson KSR, Cobbold PH. Phorbol ester and sperm activate mouse oocytes by inducing sustained oscillations in cell Ca^{2+}. Nature 1985; 316:541-542.
26. Woods NM, Cuthbertson KSR, Cobbold PH. Repetitive transient rises in cytoplasmic free calcium in hormone-stimulated hepatocytes. Nature 1986; 319:600-602.
27. Meyer T, Stryer L. Molecular model for receptor-stimulated calcium spiking. Proc Natl Acad Sci USA 1988; 85:5051-5055.
28. Schuster S, Marhl M, Höfer T. Modelling of simple and complex calcium oscillations. From single-cell responses to intercellular signalling. Eur J Biochem 2002; 269:1333-1355.
29. Falcke M. Reading the patterns in living cells - the physics of Ca^{2+} signaling. Adv Phys 2004; 53:255-440.
30. Klipp E, Heinrich R, Holzhütter HG. Prediction of temporal gene expression. Metabolic opimization by re-distribution of enzyme activities. Eur J Biochem 2002; 269:5406-5413.
31. Varner J, Ramkrishna D. Metabolic engineering from a cybernetic perspective. 1. Theoretical preliminaries. Biotechnol Prog 1999; 15:407-425.
32. Zaslaver A, Mayo AE, Rosenberg R et al. Just-in-time transcription program in metabolic pathways. Nature Genet 2004; 36:486-491.
33. Heinrich R, Rapoport TA. A linear steady-state treatment of enzymatic chains. General properties, control and effector strength. Eur J Biochem 1974; 42:89-95.
34. Schuster S, Hilgetag C. On elementary flux modes in biochemical reaction systems at steady state. J Biol Syst 1994; 2:165-182.
35. Schuster S, Dandekar T, Fell DA. Detection of elementary flux modes in biochemical networks: A promising tool for pathway analysis and metabolic engineering. Trends Biotechnol 1999; 17:53-60.
36. Schuster S, Hilgetag C, Woods JH et al. Reaction routes in biochemical reaction systems: Algebraic properties, validated calculation procedure and example from nucleotide metabolism. J Math Biol 2002; 45:1530-181.
37. Schilling CH, Letscher D, Palsson BO. Theory for the systemic definition of metabolic pathways and their use in interpreting metabolic function from a pathway-oriented perspective. J theor Biol 2000; 203:229-248.
38. Fischer E, Sauer U. A novel metabolic cycle catalyzes glucose oxidation and anaplerosis in hungry Escherichia coli. J Biol Chem 2003; 278:46446-46451.
39. Liao JC, Hou S-Y, Chao Y-P. Pathway analysis, engineering, and physiological considerations for redirecting central metabolism. Biotechn Bioeng 1996; 52:129-140.
40. Goldbeter A, Dupont G, Berridge MJ. Minimal model for signal-induced Ca^{2+} oscillations and for their frequency encoding through protein phosphorylation. Proc Natl Acad Sci USA 1990; 87:1461-1465.
41. Miller SG, Kennedy MB. Regulation of brain type II Ca^{2+}/calmodulin-dependent protein kinase by autophosphorylation: A Ca^{2+}-triggered molecular switch. Cell 1986; 44:861-870.
42. Lisman JE, Goldring MA. Feasibility of long-term storage of graded information by the Ca^{2+}/calmodulin-dependent protein kinase molecules of the postsynaptic density. Proc Natl Acad Sci USA 1988; 85:5320-5324.
43. Dupont G, Goldbeter A. Protein phosphorylation driven by intracellular calcium oscillations: A kinetic analysis. Biophys Chem 1992; 42:257-270.
44. Meyer T, Stryer L. Calcium spiking. Annu Rev Biophys Biophys Chem 1991; 20:153-174.
45. Hanson PI, Meyer T, Stryer L et al. Dual role of calmodulin in autophosphorylation of multifunctional CaM kinase may underlie decoding of calcium signals. Neuron 1994; 12:943-956.
46. Michelson S, Schulman H. CaM kinase: A model for its activation dynamics. J theor Biol 1994; 171:281-290.
47. Dosemeci A, Albers RW. A mechanism for synaptic frequency detection through autophosphorylation of CaM kinase II. Biophys J 1996; 70:2493-2501.
48. De Koninck P, Schulman H. Sensitivity of CaM kinase II to the frequency of Ca^{2+} oscillations. Science 1998; 279:227-230.
49. Dupont G, Goldbeter A. CaM kinase II as frequency decoder of Ca^{2+} oscillations. BioEssays 1998; 20:607-610.
50. Dupont G, Houart G, De Koninck P. Sensitivity of CaM kinase II to the frequency of Ca^{2+} oscillations: A simple model. Cell Calcium 2003; 34:485-497.
51. Kubota Y, Bower JM. Transient versus asymptotic dynamics of CaM kinase II: Possible roles of phosphatase. J Comput Neurosci 2001; 11:263-279.
52. Lisman JE. A mechanism for memory storage insensitive to molecular turnover: A bistable autophosphorylating kinase. Proc Natl Acad Sci USA 1985; 82:3055-3057.

SECTION IV

Mechanistic Predictions from the Analysis of Biomolecular Sequence Populations: Considering Evolution for Function Prediction

Chapter 8

Theory of Early Molecular Evolution:

Predictions and Confirmations

Edward N. Trifonov*

Abstract

A new theory of early molecular evolution is described, proceeding from original speculations to specific predictions and their confirmations. This classical cycle is then repeated generating the earliest picture of evolving Life. First, a consensus temporal order ("chronology") of appearance of amino acids and their respective codons on evolutionary scene is reconstructed on the basis of 60 different criteria, resulting in the order: G, A, D, V, P, S, E, L, T, R, I, Q, N, K, H, C, F, Y, M, W. It reveals two fundamental features: the amino acids synthesized in experiments imitating primordial conditions appeared first, while the amino acids associated with codon capture events came last. The reconstruction of codon chronology then follows based on the above consensus temporal order, supplemented by the stability and complementarity rules first suggested by M. Eigen and P. Schuster, and on earlier established processivity rule. The derived genealogy of all 64 codons suggests several important predictions that are confirmed: Gradual decay of glycine content in protein evolution; traces of the most ancient 6-residue long gly-rich and ala-rich minigenes in extant sequences; and manifestations of a fundamental binary code of protein sequences.

Introduction

Hot rocks and boiling water—that, presumably, was the "weather" on the planet Earth when 3.9 billion years ago the LIFE started (ref. 1, and references therein). It would not be fair if a skeptical reader had asked: what exactly is life? There are many answers to that question[2] though only one is needed. But it would be equally unfair to claim that the emerging life was as complex and omnipotent as today. It was surely primitive, even, perhaps, trivial, but what was it?

The one who knows what was the most primitive start is Stanley Miller who thought, in 1953, that perhaps in a primordial atmosphere a mere chemistry would take a chance. The imitation experiments[3,4] brought a spectacular result: among many other substances 10 amino acids were synthesized, half of the amino-acid repertoire of modern proteins: alanine, glycine, aspartate, valine, leucine, glutamate, serine, isoleucine, proline, and threonine (A, G, D, V, L, E, S, I, P, T). The earliest attempt of this kind, with the same thought, was the work of Löb[5] in 1913 (see also ref. 6). Analytical chemistry of that time was able to detect only one amino acid in the mixture—glycine.

Those 10 amino acids were not life yet, but a good chemical beginning, on the long way from primitive to simple, and from simple to complex. There are many dramatic stations in this journey: formation of first very small proteins, formation of the membranes and cells,

*Edward N. Trifonov—Genome Diversity Center, Institute of Evolution, University of Haifa. Mount Carmel, Haifa 31905, Israel. Email: trifonov@research.haifa.ac.il.

Discovering Biomolecular Mechanisms with Computational Biology, edited by Frank Eisenhaber.

development of replicating molecules and systems, emergence of nucleic acids, invention of the triplet code, formation of the last common ancestor, first bifurcations of the tree of Life. Each step is a mystery, and it is not clear at all what was the sequence of the events.

First Move Towards a New Theory

Let us make a jump straight to the origin of the genetic code, a pretty early stage anyway, not far from the very beginning. Within last few years my colleagues and I were lucky to have asked several very pointed, turned-to-be-right questions, and find tantalizing answers. In particular, given the earliest small proteins and nucleic acids (perhaps, RNA)—what were the very first RNA triplets (codons), and what were the amino acids they encoded? There are many speculations on that matter, including our own attempt.[7] This work, however, was not just yet another one of the speculative kind. It had an element of reconstruction of early biomolecular history, based on specific prediction that was confirmed. Such reconstruction was later expanded and turned, actually, in a vibrant theory of early molecular evolution, that suggested new predictions, followed by confirmations. The development of the theory is described in the following sections in all its logical and some technical details.

The first clue was thrown in by Thomas Bettecken, who in 1996 overviewed a group of so-called triplet expansion diseases. These neurodegenerative diseases are associated with repeating sequences located around certain genes. The repeats are of the type CUGCUGCUG or $(CUG)_n$ where n is the number of the repeats, normally 20 - 50. The repeat number all of a sudden changes to a much larger one, of the order of few hundreds, and that results in a disease. What Thomas noticed is that the most of the observed seemingly different expansions, as documented in literature, actually, correspond to the same structure. E.g., repeats $(CAG)_n$, $(GCU)_n$ and the above $(CUG)_n$, obviously, correspond to the same repeating duplex $(GCU)_n \cdot (AGC)_n$. As it turned out, the repeats $(GCU)_n$ and $(GCC)_n$ make a majority of all known triplet expansions. In other words, these two triplets are most expandable, whatever the reason. This is also confirmed in prokaryotic system.[8] This observation per se is not yet enlightening. To make the bell ring one needs another important piece of information of which we were in possession already in 1987,[9] being unaware of its explosive value. This is the $(GNN)_n$ periodicity hidden in all modern protein coding sequences. Later analysis allowed to refine the pattern to $(GCU)_n$.[10] Thus, the hidden mRNA consensus would be $(GCU)_n$, - probably, reflection of an ancient mRNA pattern $(GCU)_n$ (and, perhaps, $(GCC)_n$ as well?).

A thrilling thought then burst in: the $(GCU)_n$ and/or $(GCC)_n$, readily expanding sequences, could be indeed the first coding sequences that later evolved to the modern sequences, where the original pattern is almost lost. An obvious advantage of these sequences at that time was their exceptional ability to expand, i.e., to become longer. The relentless $(GCU)_n$ and $(GCC)_n$ repeats are still in labor—in the modern diseases.

If, thus, the GCU and GCC were the very first codons, they could only code for two amino acids. Several more amino acids should have been accommodated, probably, by single point mutations of the generic GCU and GCC triplets which gives total 15 different triplets (codons).

But which amino acids came first? One only could speculate about it, and we picked up three most natural speculations: (1) The very first amino acids were chemically simplest. (2) They would be expected to appear in the Miller's imitation mix; and (3) They would be likely to have been served by more ancient of known two classes of aminoacyl-tRNA synthetases.[11] By a consensus of these three criteria the amino acids alanine, aspartic acid, glycine, proline, serine and threonine (A, D, G, P, S, T) should have been the very first amino acids, to be served by those first 15 triplets above. Remarkably, 13 of the triplets do, indeed, encode *today* the speculated six earliest amino acids. Correspondence of the 13 predicted earliest codons to 6 predicted earliest amino acids confirms both speculations, and may be considered as a first very promising step in the possible full reconstruction of the origin and evolution of the triplet code. Encouragingly, the match between these two sets is in full agreement with present-day code. That is, being set up once, the code probably never changed.

Consensus Temporal Order of Amino Acids

There are much more than just three ways to speculate about the temporal order of appearance of the amino acids. Of these rather conflicting theories and views none is conclusive or convincing enough to become generally accepted. One could try, however, to derive a consensus of various opinions. Note that the obvious way of arguing and persuading the educated parties to an acceptable common view would be an impossible task. Speculations are difficult to challenge, if one is armed by just another speculation. There is, however, one way to arrive to the consensus—by expressing the speculations in standard way, as 20D ranking vectors, highest rank for the earliest amino acid, and averaging the vectors. Thus, all published speculations, theories, thoughts, estimates and experiments each suggesting certain amino-acid chronology are collected (currently, 60 such "opinions" are available) and expressed in form of the ranking vectors. Lead by scientific common sense and knowledge they can not be all completely out of blue. Some relevance to truth should be here and there, and simple mathematics of correlations showed that only 11 of 60 criteria did not positively correlate with the rest.[12] That is, most of the opinions, after all, resemble each other, at least distantly. After the "voting" (averaging) the commonality emerges in form of the following consensus order, the quintessence of current thought and knowledge: G (gly), A (ala), D (asp), V (val), P (pro), S (ser), E (glu), L (leu), T (thr), R (arg), I (ile), Q (gln), N (asn), H (his), K (lys), C (cys), F (phe), Y (tyr), M (met), W (trp).[12]

Apart from having a merit of consensus, the order is remarkable, in more than one way. First, the highest rank amino acids of the list (G, A, D, V, P, S, E, L, T, R) include the six earliest ones (A, D, G, P, S, T) according to our initial reconstruction with expansion triplets and three criteria of amino acid chronology (see above). What is more important—all 10 amino acids of Miller's mix (A, G, D, V, L, E, S, I, P, T) are on the top, one next to another, except for isoleucine which is 11th in the consensus chronology. That result does not change when the criteria based on the imitation experiments are excluded from the calculations.[12] A conceptual meaning of this observation is hard to underestimate. It says that all the long period of time until the triplet code reached its halfway, the amino acids incorporated in the code were just those ones that were available in the primordial chemical environment. It is amazing to see how opportunistic was Life already in the beginning. It also suggests, without even a grain of vitalism, that the opportunism is, perhaps, much more than anthropomorphic metaphor, when applied to the life processes.

Yet another very important feature of the consensus amino acid chronology is special nature of the amino acids appearing at the bottom of the list. These are of a "burglar" type—their codons appear to be actually captured from the repertoires established earlier.[13] For example, tryptophan, cysteine and tyrosine are believed to be served by codons that previously belonged to termination repertoires. Methionine is suggested to acquire its codon from isoleucine repertoire, lysine—from asparagine's. Histidine and phenylalanine are also believed to belong to that team. Here as well, very early Life demonstrates its opportunism—if codons are not available anymore—take those that are in excessive possession of others.

Reconstruction of the Origin and Evolution of the Triplet Code

It is hard to believe that such astonishingly simple and appealing temporal order of amino acids would not reflect at least some of the past realities. If it does, then the next daring task would be to try to reconstruct the early history of the triplet code, that is temporal order of codons as well.

The consensus amino acid chronology is taken as a basis for the reconstruction. It is supplemented by three most natural rules: thermostability, complementarity and processivity. The first two rules have been originally suggested by Eigen and Schuster,[14] who noted that the earliest amino acids were most likely alanine and glycine, which are the highest yield amino acids in the Miller's imitation mix. Among their respective codons the triplets GCC and GGC make the most stable complementary contact of 32 possible complementary triplet pairs. These two first codon assignments, thus, had to be introduced simultaneously, suggesting that, perhaps, all other codons would enter as complementary pairs as well. The processivity rule says that new codons are not created de novo (as, e.g., the very first codon pair GGC·GCC) but rather are simple mutational versions of the codons already present.

The result of the reconstruction is presented in Figure 1. Here the lines correspond to complementary pairs of codons. The order from left to right corresponds to the consensus temporal order of amino acids. The order from top to bottom corresponds to the temporal order of appearance of the 32 complementary codon pairs that also goes with monotonically descending order of their thermostability, experimental error bars respected (see the left-most column).

According to the reconstruction chart, the pair gly (GGC) and ala (GCC) are the very first, generic elements in the evolution of the codon table. Next in amino acid chronology are asp and val, with their complementary codons GAC and GUC. The processivity rule requires that these codons would be formed from the GGC and/or GCC. Indeed, it is realized by transitions G to A or C to U, and subsequent complementary copying (GGC → GAC → GUC, and/or GCC → GUC → GAC). Next in the amino acid order is proline. The most stable codon of its present-day repertoire is CCC. It is, obviously, derived from already present glycine codon GGC by mutation in the redundant third position, and complementary copying (GGC → GGG → CCC). This simple scheme of derivation of new codons is repeated then all the way to the bottom of the chart. When all 64 triplets are exhausted, the codon capture stage starts. The capture is rather violent. It does not follow neither the thermostability rule, nor complementarity rule. The capture cases do, however, follow the amino acid chronology and the rule of processivity.

The most striking characteristic of the reconstruction is its absolute loyalty to the most basic simple rules (thermostability, complementarity and processivity), given the consensus temporal order of amino acids. If the order is violated by swapping positions of, say, proline and methionine, the consistency of the chart is ruined. Only small changes would be allowed, within the error bars of the stability values and error limits in the rankings in the amino acid order. These are, typically, less than one rank unit.[12]

Another astounding feature is that at no step in the reconstruction any unusual codon assignment or unusual amino acid was suggested. That simply means that even at the earliest stages of the evolution of the triplet code it was largely the same as the present-day universal code, only of smaller repertoire.

The chart has several features that suggest very specific predictions.

Prediction I. Early Proteins Were Glycine-Rich

That would follow from the earliest steps of the chart, during which the composition of glycine dropped from 50% at the first step to about 33% at step 6 (considering the codons equally frequent). At later stages the glycine composition gradually dropped further to modern level of 7 - 8%. One, thus, would expect that the ancient proteins were rather glycine-rich. One more possible reason for the dominance of glycine in the early stages is high flexibility of glycine-rich chains, to ensure greater conformational diversity of the early proteins. That factor may have been interwound with the logic of consecutive introduction of new codons and amino acids. The glycine-richness would be a strong prediction if we only would have the specimen of the ancient protein, to check the amino acid composition of the ancient sample and compare it with modern proteins. The problem is that the oldest fossils date back only one billion years and are, thus, rather "young". Those modern proteins that are thought to have longest evolutionary history, are not a good sample since all proteins, of all modern organisms, are of the same age. The same time passed for all of them since whatever common time reference. Thus, the full sequences of the ancient proteins proper are not available. However, small patches of the ancient patterns seem to survive in some modern sequences. For example, functionally similar sequences from eukaryotes and prokaryotes

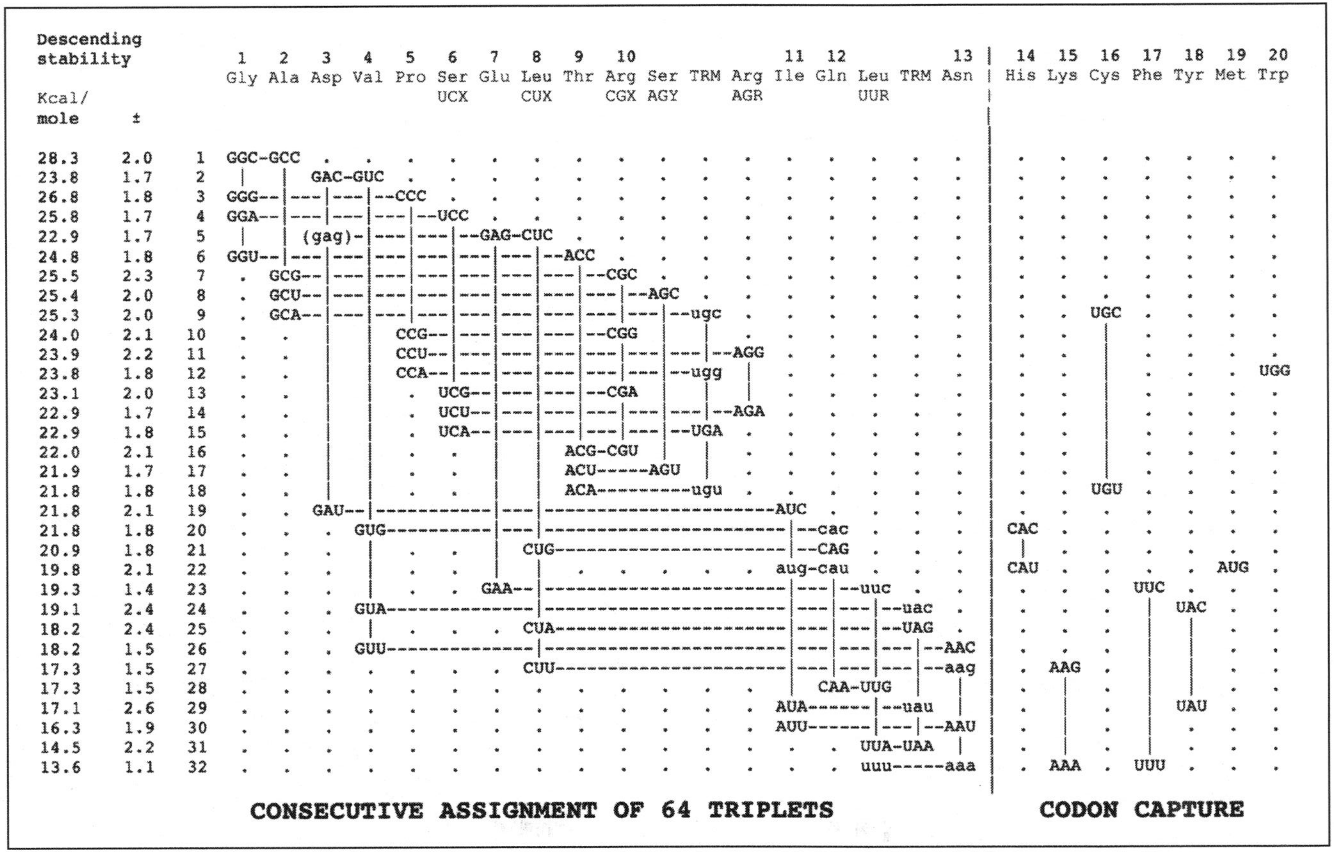

```
Descending
stability        1   2   3   4   5   6   7   8   9   10              11  12          13 | 14  15  16  17  18  19  20
                Gly Ala Asp Val Pro Ser Glu Leu Thr Arg Ser TRM Arg Ile Gln Leu TRM Asn | His Lys Cys Phe Tyr Met Trp
Kcal/                               UCX     CUX     CGX AGY     AGR         UUR         |
mole     ±                                                                              |

28.3    2.0   1 GGC-GCC .   .   .   .   .   .   .   .   .   .   .   .   .   .   .   .   |  .   .   .   .   .   .   .
23.8    1.7   2  |   |  GAC-GUC .   .   .   .   .   .   .   .   .   .   .   .   .   .   |  .   .   .   .   .   .   .
26.8    1.8   3 GGG--|---|---|--CCC .   .   .   .   .   .   .   .   .   .   .   .   .   |  .   .   .   .   .   .   .
25.8    1.7   4 GGA--|---|---|---|--UCC .   .   .   .   .   .   .   .   .   .   .   .   |  .   .   .   .   .   .   .
22.9    1.7   5  |   | (gag)-|---|---|--GAG-CUC .   .   .   .   .   .   .   .   .   .   |  .   .   .   .   .   .   .
24.8    1.8   6 GGU--|---|---|---|---|---|---|--ACC .   .   .   .   .   .   .   .   .   |  .   .   .   .   .   .   .
25.5    2.3   7  .  GCG--|---|---|---|---|---|---|--CGC .   .   .   .   .   .   .   .   |  .   .   .   .   .   .   .
25.4    2.0   8  .  GCU--|---|---|---|---|---|---|---|--AGC .   .   .   .   .   .   .   |  .   .   .   .   .   .   .
25.3    2.0   9  .  GCA--|---|---|---|---|---|---|---|---|--ugc .   .   .   .   .   .   |  .   .  UGC  .   .   .   .
24.0    2.1  10  .   .   |   |  CCG--|---|---|---|--CGG |   |   .   .   .   .   .   .   |  .   .   |   .   .   .   .
23.9    2.2  11  .   .   |   |  CCU--|---|---|---|---|---|---|--AGG .   .   .   .   .   |  .   .   |   .   .   .   .
23.8    1.8  12  .   .   |   |  CCA--|---|---|---|---|---|--ugg |   .   .   .   .   .   |  .   .   |   .   .   .  UGG
23.1    2.0  13  .   .   |   |   .  UCG--|---|---|--CGA |   |   |   .   .   .   .   .   |  .   .   |   .   .   .   .
22.9    1.7  14  .   .   |   |   .  UCU--|---|---|---|---|---|--AGA .   .   .   .   .   |  .   .   |   .   .   .   .
22.9    1.8  15  .   .   |   |   .  UCA--|---|---|---|---|--UGA .   .   .   .   .   .   |  .   .   |   .   .   .   .
22.0    2.1  16  .   .   |   |   .   .   |   |  ACG-CGU |   |   .   .   .   .   .   .   |  .   .   |   .   .   .   .
21.9    1.7  17  .   .   |   |   .   .   |   |  ACU-----AGU |   .   .   .   .   .   .   |  .   .   |   .   .   .   .
21.8    1.8  18  .   .   |   |   .   .   |   |  ACA---------ugu .   .   .   .   .   .   |  .   .  UGU  .   .   .   .
21.8    2.1  19  .   .  GAU--|-----------|---|---------------------AUC .   .   .   .   |  .   .   .   .   .   .   .
21.8    1.8  20  .   .   .  GUG----------|---|-----------------------|--cac .   .   .   | CAC  .   .   .   .   .   .
20.9    1.8  21  .   .   .   |   .   .   |  CUG----------------------|--CAG .   .   .   |  |   .   .   .   .   .   .
19.8    2.1  22  .   .   .   |   .   .   |   |   .   .   .   .   .  aug-cau .   .   .   | CAU  .   .   .   .  AUG  .
19.3    1.4  23  .   .   .   |   .   .  GAA--|-----------------------|---|--uuc .   .   |  .   .   .  UUC  .   .   .
19.1    2.4  24  .   .   .  GUA--------------|-----------------------|---|---|--uac .   |  .   .   .   |  UAC  .   .
18.2    2.4  25  .   .   .   |   .   .   .  CUA----------------------|---|---|--UAG .   |  .   .   .   |   |   .   .
18.2    1.5  26  .   .   .  GUU--------------|-----------------------|---|---|---|--AAC |  .   .   .   |   |   .   .
17.3    1.5  27  .   .   .   .   .   .   .  CUU----------------------|---|---|---|--aag |  .  AAG  .   |   |   .   .
17.3    1.5  28  .   .   .   .   .   .   .   .   .   .   .   .   .   |  CAA-UUG  |   |   |  .   |   .   |   |   .   .
17.1    2.6  29  .   .   .   .   .   .   .   .   .   .   .   .   .  AUA------|--uau  |   |  .   |   .   |  UAU  .   .
16.3    1.9  30  .   .   .   .   .   .   .   .   .   .   .   .   .  AUU------|---|--AAU   |  .   |   .   |   .   .   .
14.5    2.2  31  .   .   .   .   .   .   .   .   .   .   .   .   .   .   .  UUA-UAA  |   |  .   |   .   |   .   .   .
13.6    1.1  32  .   .   .   .   .   .   .   .   .   .   .   .   .   .   .  uuu-----aaa   |  .  AAA  .  UUU  .   .   .

                 CONSECUTIVE ASSIGNMENT OF 64 TRIPLETS                                       CODON CAPTURE
```

Figure 1. Origin and evolution of the triplet code. The top line—temporal order of amino acids. Lines 1 to 32—temporal order of complementary triplet pairs. The column on the left gives the descending order of thermostability[12] of the triplet pairs.

(time of separation about 3.5 billion years) do show sequence similarities. Collecting those patches that are identical in the pairs compared, one can calculate their amino acid composition. Such calculation[15] demonstrates that the surviving pieces of ancient protein sequences are, indeed, glycine-rich (14%).

Moreover, comparing in similar way proteins of Eubacteria, Archaea, Protoctista, Fungi, Plants and Animals, one can derive the respective values of glycine content for every kingdom-to-kingdom pair. That results in the reconstruction of rooted tree of 6 major kingdoms, fully consistent with current knowledge.[15] The prediction, thus, is well confirmed.

Prediction II. The Earliest Protein Sequences Were a Mosaic of Two Independent Alphabets

The genealogy of the triplet code presented in Figure 1 contains a wealth of information to explore. It says, for example, that the very first genes could be $(GGC)_n$ and, complementary, $(GCC)_n$ repeats, encoding monotonous chains of two types consisting either of glycine residues only, or of alanine only. That view is supported by the expandability of the repeats (see above). Later replacements, according to the genealogy, diluted this domination of glycine and alanine, but, perhaps, at least in the earliest evolution the glycine-rich and the alanine-rich pieces remained. How long were these first protein molecules? Just on the basis of solubility of these oligoglycines and oligoalanines one can say that they were not longer than 7-8 residues. The longer ones would aggregate, thus, excluding themselves from all molecular interactions required by the early life. Today protein chains are as long as typically 100 – 200 amino acids. They are soluble due to a balanced proportion of polar residues. Those original short pieces, after acquiring first charged amino acids, aspartic (D) and glutamic (E) acids, could fuse, making larger molecules, well soluble due to the added charges. In this case the alanine-rich and glycine-rich sections would frequently alternate. Inspection of the chart reveals one additional feature to describe the hypothetical short sections. The RNA strand originally encoding only glycines first acquired additional codon GAC, for aspartate, then GAG, for glutamate, CGC for arginine, and so on, keeping always the purine nucleotide, G or A, in the middle of the codons. That is, the "glycine strand" during all the evolution of the triplet code carried only codons with middle purines. Similarly, the complementary "alanine strand" carried only codons with middle pyrimidines. That splits 20 amino acids in two practically independent alphabets, Gly-alphabet (G, D, E, R, S, Q, N, K, H, C, Y, W) and Ala-alphabet (A, P, S, L, T, I, F, M). Only one amino acid, serine, is shared. Thus, both before and after fusion of the original minigenes, they represented Gly-strand segments, encoding Gly-alphabet, and Ala-strand segments, encoding Ala-alphabet. The mutations, all in redundant third positions, have lead to complementary changes in the first positions, while the middle positions stayed unchanged, thus, keeping alphabet identity of the original minigenes and of their later mutational versions unchanged. The prediction follows: one may hope, as we did, that such hypothetical ancient alternating motif Ala-alphabet/Gly-alphabet may still be recognized in modern protein sequences, despite billions of years of mutational changes. To our delight this prediction was, indeed, confirmed by massive analysis of completely sequenced full sets of proteins of 23 different bacteria.[16] The earliest unit size, the minigene, was determined as well. It was 6 amino acid residues, fitting well to the solubility limit estimate.

Prediction III. Fundamental Binary Code of Protein Sequences

The two alphabets, as outlined above, had to be maintained in pure form during all the time until the time-wise uncertain stage when the original double-stranded coding molecules switched to coding in only one strand, as today. All sequence patterns that painfully developed during that time to their best performance, were transferred to the single-strand coding form, so that from this moment on, the obligatory pressure of alphabet purity was lost. It is worth

		A	F	I	L	M	P	T	V	C	D	E	G	H	K	N	Q	R	W	Y
	A						1	1					1							4
	F																			
	I				1	1			3											
Ala	L			1		3			1											
alphabet	M			1	3				1											
	P	1																		
	T	1																		
	V			3	1	1														
	C																			
	D											3				2	1			
	E										3					1	2			
	G	1																		
Gly	H															2	3	1		
alphabet	K															1		2		
	N										2	1		2	1					
	Q										1	2		3				1		
	R													1	2		1		1	
	W																	1		2
	Y	4																	2	

Figure 2. PAM-120 matrix, as presented in reference 17, is rearranged here in form emphasizing two alphabets. The strongest matrix elements (in bits) are shown, as in reference 17.

noting, however, that since the most frequent type of point mutations is transition (purine-to-purine and pyrimidine-to-pyrimidine) the ancient patterns expressed as sequences in two-letter alphabets would continue to be maintained at least for some time. Hopefully, the patterns may still be recognizable in the modern sequences. Descendants of the same original pattern, expressed in traditional 20-letter alphabet, may appear rather different, while their binary presentations would suggest close relatedness. The prediction is: the nonmatching residues in related protein sequences (20-letter form) should rather match when the sequences are presented in the binary form.

There is a simple and straightforward way to check the prediction. Large collections of the related sequences have been already compared for the purpose of finding out which residues more frequently change to which. The changes were tabulated in form of well known PAM and BLOSSUM matrices broadly used in the calculations of sequence relatedness. All what remains to be done is to simply check whether these matrices, actually, demonstrate mostly the changes within the same alphabet.

In Figure 2, the PAM-120 matrix[17] that shows only most frequent replacements, is reorganized in such a way that all amino acids are ordered in the two separate groups, according to the two generic Ala- and Gly-alphabets. The resulting new matrix clearly shows two separate boxes, demonstrating the expected within-the-alphabet replacements. Similarly, in Figure 3, the BLOSSUM matrix[18] after such rearrangement shows the same strong effect: the replacements occur with dominant preference to the amino acids of the same alphabet, either Ala-type or Gly-type, as expected.

This observation, predicted on the basis of the chart of codons presented in Figure 1, has important consequences. First, it would allow to outline the most ancient sequence patterns and to trace very early evolutionary moves. Second, more to earth, it would help to identify and classify related sequences in those cases when the 20-letter comparisons fail to detect the relatedness.

		A	F	I	L	M	P	T	V	C	D	E	G	H	K	N	Q	R	W	Y
	A																			
	F																		1	3
	I				2	1			3											
Ala	L			2		2			1											
alphabet	M			1	2				1											
	P																			
	T																			
	V			3	1	1														
	C																			
	D											2				1				
	E										2				1		2			
	G																			
Gly	H															1				2
alphabet	K											1					1	2		
	N										1			1						
	Q											2			1			1		
	R														2		1			
	W		1																	2
	Y		3											2					2	

Figure 3. BLOSSUM matrix[18] rearranged here in the same form as above in Figure 2. The strongest matrix elements (in bits) are shown, as in original.

Prediction IV. Domestication of Life?

From the very onset of Life (3.9 billion years ago), which was presumably purely chemical, as in the imitation experiments of S. Miller, until the first major separation of already developed life forms, eukaryotes and prokaryotes (3.5 billion years back), good 400 million years passed. Within that span the triplet code emerged and evolved to its close-to-complete form. The reconstruction of the evolution of the code, presented in Figure 1, originally a sophisticated speculation, is now fortified by several confirmed predictions, that upgrades it to a theory. The theory provides the deepest top-down view into the Life's past. The time gap between first abiotically synthesized biologically relevant substances, and the first codons, GGC and GCC, is still to be filled by unknown specifics, and that would be, perhaps, the most difficult part of reconstruction of the origin of Life. Before that, however, one could try to explore and, possibly, to further expand the picture provided by the above reconstruction of the early evolution of the codons.

The very first codons, GGC and GCC, can only deliver monotonous unchanging oligomers: $(GGC)_n$, $(GCC)_n$, $(gly)_n$ and $(ala)_n$. But as soon as two more codons are mutationally introduced, GAC and GUC (see Fig. 1), for asp (D) and val (V), respectively, the homo-oligomers would turn into hetero-oligomers, with a variety of sequence possibilities. Some of the variants under pressure of survival would do better than original monotonous GGGGGG and AAAAAA. It appears then, that the very border between pure chemistry (constancy only) and life (constancy and variability) passes between the first and second lines of the codon genealogy. This gives another good reason to try to explore *experimentally* this earliest transition from nonliving to living matter.

The hexamers $(GGC)_6$ and $(GCC)_6$, together with peptides $(ala)_6$ and $(gly)_6$, may be considered as sort of composite replicator that may work as such, supplemented of course with monomer units GGC, GCC, ala and gly as building material, some source of energy, mineral catalyzers, and other conditions, still to find (http://research.haifa.ac.il/~genom/Trifonov/Origin/index.htm). To get even mere replication of some components in the mixture would be a major achievement. The prediction is that it may work! One day, driven by logic of the developments of the new theory, somebody will pick up the idea, find necessary support and set the fire of the domesticated Life.

Linking the Codon Chronology with Other Events of Early Evolution

The very first "proteins", according to the confirmed prediction based on the chart of the codons (Fig. 1) were 6 residue long oligopeptides. After fusion of their respective minigenes longer chains could be formed. It is at this stage when the first structural features would appear. In particular, the polymer chain statistics suggests that at certain optimal length of the chain, 25 - 30 amino acid residues, its ends would frequently engage in contacts, making closed loops.[19] The contacts fortified by van der Waals interactions between the residues at the ends would render the loops stability, an obvious selective advantage in the early evolution of proteins.[20] It appears that the 25 - 30 residue loop modules are the major elements of modern proteins, and many of them can be traced far back to their ancestry.[16,20,21] Is that important stage in the evolution of proteins reflected somehow in the chart of the evolution of the triplet code (Fig. 1)? The initial 18 or so base-pair long RNA duplexes with the advance of the repertoire of codons and amino acids were growing, either due to triplet expansion (slippage) or by minigene fusion. The encoded protein chains grew as well. Since the size of 25-30 amino acids was optimal, further elongation of the encoded sequences should have stopped at this stage of evolution. One way to ensure the length limitation would be to introduce either initiation or stop codons. RNA strands beyond these codons would be excessive, in both coding strands. That is, RNA duplexes as well would be limited to the length about 75 - 90 base pairs. Inspection of Figure 1 reveals that the first four termination codons of the UGX family appeared at steps 9 to 18. This puts tentative limits to the loop closure stage in the protein evolution—soon after incorporation of arginine into the triplet code, but before acquirement of isoleucine. The above reasoning is, of course, a speculation. Future developments of the theory will show how fruitful this speculation is.

The next stage of the protein evolution,[16,22] as originally suggested by studies on the sizes of modern proteins[23,24] is fusion of the 75 - 90 base pair long genes to the size 350 - 450 base pairs, optimal for DNA circularization.[25] This size corresponds to the typical protein fold sizes, 100 - 150 amino acid residues. The RNA fragments encoding the standard closed loop proteins (continuing the speculation above) could increase their length by fusion only after appearance of the first initiation triplets, to ensure the initiation within the molecule rather than just at its ends. That would allow the fusion of the short RNA duplexes into longer ones carrying several small genes, each one with its own initiation and termination signals (steps 19 to 31 of the chart). The next stage would be fusion of the short coding sequences within the long RNA duplexes. The short 25-30 codon genes would require more initiation and termination triplets than the 100 - 150 codon genes. Respectively, the fusion of the short coding sequences can be projected to the codon capture stage when the terminators UGX and UAX yielded to cysteine and tyrosine, and AUG became practically the sole initiation signal (Fig. 1). The transition from RNA coding (steps 1 to 32) to DNA coding, presumably, occured at the onset of the codon capture stage, with appearance of histidine and lysine.

Many more early molecular events can be tentatively placed within the codon evolution chart. It opens a whole new fertile field for further speculations and predictions towards detailed reconstruction of the Life's distant past.

References

1. Di Giulio M. The universal ancestor and the ancestor of bacteria were hyperthermophiles. J Molec Evol 2003; 57:721-730.
2. Barbieri M. The Organic Codes. An introduction to semantic biology. Cambridge University Press, 2003.
3. Miller SL. A production of amino acids under possible primitive earth conditions. Science 1953; 117:528-529.
4. Miller SL. Which organic compounds could have occurred on the prebiotic Earth? Cold Spr Harb Symp Quant Biol 1987; 52:17-27.
5. Löb W. Über das Verhalten des formamids unter der wirkung der stillen entladung: Ein betrag zur frage der stickstoff-assimilation. Ber 1913; 46:684-697.
6. Yockey HP. Walther Löb, Stanley L. Miller and prebiotic "building blocks" in the silent electrical discharge. Persp Biol Med 1997; 41:125-131.
7. Trifonov EN, Bettecken T. Sequence fossils, triplet expansion, and reconstruction of earliest codons. Gene 1997; 205:1-6.
8. Ohshima K, Kang S, Wells RD. CTG triplet repeats from human hereditary diseases are dominant genetic expansion products in Escherichia coli. J Biol Chem 1996; 271:1853-1856.
9. Trifonov EN. Translation framing code and frame-monitoring mechanism as suggested by the analysis of mRNA and 16S rRNA nucleotide sequences. J Molec Biol 1987; 194:643-652.
10. Lagunez-Otero J, Trifonov EN. mRNA periodical infrastructure complementary to the proof-reading site in the ribosome. J Biomolec Str Dyn 1992; 10:455-464.
11. Eriani G, Delarue M, Poch O et al. Partition of tRNA synthetases into two classes based on mutually exclusive sets of sequence motifs. Nature 1990; 347:203-206.
12. Trifonov EN. The triplet code frm first principles. J Biomol Struct Dyn 2004; 22:1-11.
13. Osawa S, Jukes TS, Watanabe K et al. Recent evidence for evolution of the genetic code. Microb Rev 1992; 56:229-264.
14. Eigen M, Schuster P. The hypercycle. A principle of natural self-organization. Part C: The realistic hypercycle. Naturwissenschaften 1978; 65:341-369.
15. Trifonov EN. Glycine clock: Eubacteria first, Archaea next, protoctista, Fungi, Planta and Animalia at last. Gene Therapy Mol Biol 1999; 4:313-322.
16. Trifonov EN, Kirzhner A, Kirzhner VM et al. Distinct stages of protein evolution as suggested by protein sequence analysis. J Mol Evol 2001; 53:394-401.
17. Altschul SF. Amino acid substitution matrices from an information theoretic perspective. J Mol Biol 1991; 219:555-565.
18. Henikoff S, Henikoff JG. Amino acid substitution matrices from protein blocks. Proc Natl Acad Sci USA 1992; 89:10915-10919.
19. Berezovsky IN, Grosberg AY, Trifonov EN. Closed loops of nearly standard size: Common basic element of protein structure. FEBS Letters 2000; 466:283-286.
20. Berezovsky IN, Kirzhner VM, Kirzhner A et al. Protein sequences yield a proteomic code. J Biomol Struct Dyn 2003; 21:317-325.
21. Berezovsky IN, Kirzhner A, Kirzhner VM et al. Spelling protein structure. J Biomol Struct Dyn 2003; 21:327-339.
22. Trifonov EN. Segmented genome: Elementary units of genome structure. Russian J Genetics 2002; 38:659-663.
23. Berman AL, Kolker E, Trifonov EN. Underlying order in protein sequence organization. Proc Natl Acad Sci USA 1994; 91:4044-4047.
24. Kolker E, Tjaden BC, Hubley R et al. Spectral analysis of distributions: Finding periodic components in eukaryotic enzyme length data. OMICS: Journal Integr Biol 2002; 6:123-130.
25. Shore D, Langowski J, Baldwin RL. DNA flexibility studied by covalent closure of short fragments into circles. Proc Natl Acad Sci USA 1981; 78:4833-4837.

CHAPTER 9

Hitchhiking Mapping:
Limitations and Potential for the Identification of Ecologically Important Genes

Christian Schlötterer*

Abstract

A recent series of publications demonstrated that identification of genomic regions subjected to positive selection (hitchhiking mapping) is possible and could be applied in an ecological context. This review focuses on the use of microsatellite markers in genome scans for the identification of beneficial mutations. The pitfalls and potential of the lnRθ test statistic are discussed as well as different approaches for the identification of the molecular change(s) underlying an observed selective sweep.

Introduction

Ecological genomics encapsulates a recent trend to apply high-throughput genomic tools to questions in ecology and evolution.[1] Progress in genomics technology has shifted the focus from the analysis of a small number of candidate genes to multiple genomic regions in several populations. While this research area is still in its infancy, already a considerable number of studies have demonstrated the enormous potential of multilocus approaches for the identification of genomic regions bearing ecologically important loci/alleles.[2,3]

This approach, which has been termed hitchhiking mapping[3,4] or selection mapping,[5] relies on a very simple population genetic principle. Theory predicts that a beneficial mutation is either lost quickly due to genetic drift or becomes fixed in the population. Importantly, not only the beneficial mutation increases in frequency, but also other, neutral variants linked to the target of selection (hitchhiking[6]). Thus, as a consequence of the spread of a beneficial mutation in a population, the allele frequency spectrum is significantly distorted from neutral expectations in a genomic region around the target of selection. Population genetics has devised a range of different approaches to use this change in allele frequency spectrum for the identification of past episodes of nonneutral evolution.[7,8]

Many of these classic neutrality tests, such as Tajima's D,[9] are affected by demographic effects, such as bottlenecks and admixture, preventing the use of a nominal P-value for the identification of selected loci.[10] When a large number of loci are surveyed, however, it is possible to build an empirical distribution of any test statistic used to quantify the distortion in allele frequency.[11] Rather than relying on a nominal P-value, genomic regions possibly subjected to selective forces are then identified as loci in the tails of the empirical distribution. While this approach eliminates the effects of demographic events, its disadvantage is that even in the absence of selection, genomic regions will be (falsely) identified as targets of selection, due to their location in the tail of the distribution.

*Christian Schlötterer—Institut für Tierzucht und Genetik, Veterinärmedizinische Universität Wien, A-1210 Wien, Austria, Europe. Email: Christian.schloetterer@vu-wien.ac.at

Discovering Biomolecular Mechanisms with Computational Biology, edited by Frank Eisenhaber.

Here, statistical approaches of hitchhiking mapping using microsatellite variation will be introduced. The general limitations of the hitchhiking mapping approach will be demonstrated using microsatellite data and possible approaches to overcome them will be discussed.

Microsatellites—A Widely Used Genetic Marker

Since their introduction in 1989[12-14] microsatellites have developed into one of the most commonly used genetic markers.[15] A microsatellite consists of a tandem repetition of one repeat motif, such as $(GT)_n$ or $(GAC)_n$. Due to DNA replication slippage, a mutation process specific to tandemly repeated DNA, the copy number of the repeat units changes at a high rate (up to 10^{-2}).[16,17] Using locus-specific PCR primers flanking a microsatellite, the variation in repeat number can be easily detected.

Most microsatellite mutations encompass either the gain or loss of a single repeat unit, but larger changes in repeat number have also been described. Nevertheless, the mutation process of microsatellites can be well-approximated by the stepwise mutation model, which was originally introduced to describe the evolution of proteins.[18] The interpretation of microsatellite variability data, in particular the comparison across loci, is significantly complicated by pronounced locus specific mutation rates.[19,20]

The lnRθ Statistic

The Concept

One of the possible consequences of a selective sweep is a reduction in variability at a genomic region subjected to a selective sweep. Thus, genome scans for targets of selection can aim for the identification of genomic regions bearing microsatellites that have less variability than expected under neutrality. The large variation in microsatellite mutation rates, however, significantly complicates the interpretation of variability patterns, as it is not possible to distinguish whether a locus has low levels of variability due to a selective sweep or a low mutation rate. In an attempt to overcome the problem of locus specific mutation rates, lnRθ has been proposed as a means of identifying microsatellite loci, which show a more pronounced reduction in variability.[21,22] Rather than analyzing microsatellite variability in one population only, the lnRθ statistic requires polymorphism data from two populations. Assuming that the microsatellite mutation rate does not differ among populations, the expectation for lnRθ is the same for all microsatellite loci, independent of the mutation rate.

$$\ln\left[\mathrm{E}(R\theta)\right] = \ln\left[\mathrm{E}\left(\frac{\frac{1}{2}\theta_{Pop1}}{\frac{1}{2}\theta_{Pop2}}\right)\right] = \ln\left[\mathrm{E}\left(\frac{\left(2N_{e_{Pop1}}\mu\right)}{\left(2N_{e_{Pop2}}\mu\right)}\right)\right] \cong \ln\left[\frac{\mathrm{E}\left(2N_{e_{Pop1}}\mu\right)}{\mathrm{E}\left(2N_{e_{Pop2}}\mu\right)}\right] \quad \text{(ref. 23)} \quad (1)$$

Two different estimators for θ could be used: variance in repeat number (V) and gene diversity (H, expected heterozygosity):

$$\theta = 2V = 4N_e\mu = \frac{2}{n-1}\sum_{i=1}^{n}\left(x_i - \bar{x}\right)^2 \quad \text{(ref. 23)} \quad (2)$$

x_i is the repeat number of allele i.

$$\theta = \left(\left(\frac{1}{1-H_{Pop1}}\right)^2 - 1\right)\frac{1}{8\mu} \quad \text{(ref. 18)} \quad (3)$$

θ estimates based on the variance in repeat number have a larger variance and are thus less sensitive to identify loci subjected to a selective sweep than θ estimates based on gene diversity.

Furthermore, the variance in repeat number is strongly affected by indel mutations in the flanking sequence of a microsatellite, but gene diversity is not.[24] On the other hand, θ estimates based on gene diversity are biased and underestimate θ, in particular, for large θ-values.[25] Nevertheless, the identification of selective sweeps based on gene diversity seems not to be affected by this bias.[25]

Identification of Selected Loci

Computer simulations indicate that lnRθ values are, to a very good approximation, normally distributed.[21,22] Given that the mean and standard deviation depends on the choice of populations, it is advised to standardize the distribution of lnRθ values. The preferred method of standardization is to use a set of neutral loci from the same populations in order to obtain the mean and standard deviation of the distribution of lnRθ values. After standardization of the lnRθ values, the probability that a single locus deviates from neutral expectations can be obtained from the density function of a standard normal distribution. Note that X-linked and autosomal loci have different effective population sizes. Thus, the standardization is complicated if the loci used for standardization are located on different chromosomes than the loci, which should be analyzed (see also Box 1).

In the absence of a neutral set of loci, for studies based on a large number of loci it is possible to use the full data set for standardization as long as only a small fraction of loci is affected by selection. In the case that a considerable fraction of the analyzed loci has been subjected to directional selection, only a bad fit to the normal distribution is obtained. If the removal of outlier loci improves the fit to a normal distribution, this can be regarded as very strong evidence for a nonneutral evolution of these loci (see also Box 1).

Influence of Demography and Choice of Populations

To a very good approximation the distribution of lnRθ values remains normal, even if one of the populations has been exposed to demographic events, such as bottlenecks or admixture from a third population.[21,22] While the power to detect deviations from neutrality is affected by demographic events, the overall applicability of the lnRθ statistic to populations with an unknown demographic past is a particularly strong advantage compared to many other neutrality tests. The application of the lnRθ statistic is not necessarily confined to the analysis of two distinct populations. It is also possible to split members of a population according to phenotype. Thus, a population can be grouped into diseased and healthy individuals or insecticide resistant and sensitive individuals. If one of the traits has emerged (and/or increased in frequency) recently, then the lnRθ statistic could be used to identify markers associated with the trait.

Mapping the Target of Selection

Genome scans for the identification of selected regions are usually carried out at a low density. The power to detect the signature of selection decreases with increasing genetic distance from the target of selection.[26,27] Thus, depending on the density of the markers, a considerable fraction of the genomic regions subjected to directional selection may be missed. Strong selection and low recombination rates will require a lower marker density than weak selection and high recombination. The drawback of strong selection and low recombination is that the identification of the target of selection is more difficult:[26,27] as a large genomic region is affected by a sweep, the microsatellite identified as a nonneutrally evolving locus, may be located several kilobases (kb) away from the actual target of selection.

Nevertheless, once a candidate region has been targeted in a primary screen, it is possible to analyze additional microsatellites mapping to the identified region. Recently, Harr et al (2002) typed several microsatellites mapping to a candidate region for a selective sweep. To map the target of selection the authors relied on a recently developed analytical framework.[27] Assuming that a selective sweep has just been completed, it is possible to predict the reduction in variability at a linked microsatellite. If more than a single microsatellite is affected by the selective sweep, it is possible to predict the position of the selected site.[4]

$$x = \frac{-4m \ln\left[-\left(\frac{V_{Pop1\_loc1}}{V_{Pop2\_loc1} - V_{Pop1\_loc1}}\right)\right] - knr \ln\left[-\left(\frac{V_{Pop1\_loc1}}{V_{Pop2\_loc1} - V_{Pop1\_loc1}}\right)\right] + 4m \ln\left[-\left(\frac{V_{Pop1\_loc1}}{V_{Pop2\_loc2} - V_{Pop1\_loc2}}\right)\right]}{nr \ln\left[-\left(\frac{V_{Pop1\_loc1}}{V_{Pop2\_loc1} - V_{Pop1\_loc1}}\right)\right] - nr \ln\left[-\left(\frac{V_{Pop1\_loc2}}{V_{nPop2\_loc2} - V_{Pop1\_loc2}}\right)\right]}$$

x is the distance of the microsatellite with the strongest reduction in variability (locus 1), m is the microsatellite mutation rate and n is the number of neutral microsatellite alleles, r is the recombination rate per base pair and k is the distance between the two microsatellite loci (in bp). Pop1 is the population representing the ancestral variation, while Pop2 is the population, which experienced the selective sweep. A slight modification of this equation could be used for the case in which the selected site is assumed to be located between the two characterized microsatellite loci.

Nevertheless, similar to other approaches attempting to map the target of selection using DNA sequence polymorphism data,[28] the greatest weakness of this microsatellite based method is that it relies on the expected reduction in variability. As shown in the next paragraph, the actual realization of a selective sweep may differ substantially from the expectation, thus the map position determined by the expectation for a selective sweep may not always be very accurate.

Large Variation in the Realization of Selective Sweeps

The power of hitchhiking mapping depends to a large degree on the extent to which the signal of a selective sweep is reflected in the surveyed genomic region. Figure 1. shows the variance in repeat number averaged over 200 independent simulation runs. In the middle of the chromosome a selective sweep occurred, which results in a pronounced reduction in variability. This genomic region with the marked loss in variability could be easily distinguished from the flanking loci, which are not affected by selection (Fig. 1).

For comparison, the average variance in repeat number around the selected site is plotted together with three individual realizations of the selective sweep (Fig. 2). In simulation run 1 a very broad window of reduced variability is obtained, which would not permit mapping the selected site with moderate precision. Run 2 also results in a broad window of reduced variability, but the locus with the most pronounced reduction in variability coincides with the target of selection, thus this run would have resulted in an acceptable map position. Run 3 in contrast, also shows several loci with reduced variability, but the locus with the most pronounced reduction in variability does not correspond to the target of selection. Thus, a wrong genomic region would have been implied as the target of selection. Overall, the computer simulations clearly indicate that a single realization of a selective sweep may not always result in the accurate determination of the position of the selected site.

Increasing the Precision of Hitchhiking Mapping

Given that the realization of a single selective sweep has a large variance, an improved mapping strategy would rely on multiple selective sweeps rather than on a single sweep. Figure 3 indicates the mean over 10 independent computer simulations. Each of the five means quite accurately maps the position of the selective sweep, strongly supporting the idea that multiple independent realizations of the same sweep are clearly superior for hitchhiking mapping than single sweeps. Independent selective sweeps could be obtained if different, geographically isolated populations are exposed to the same selective force. Clines provide an

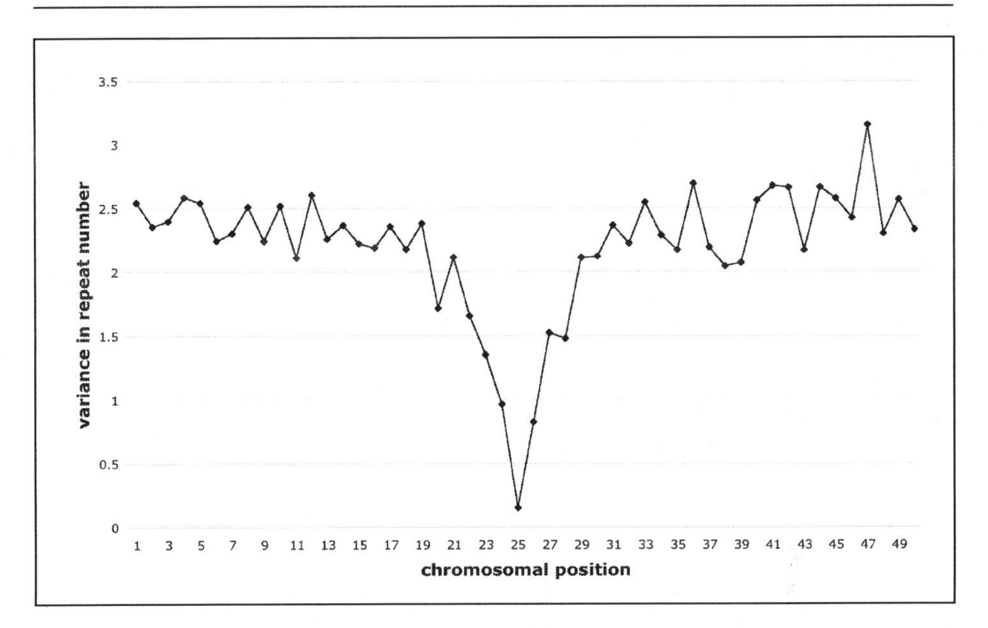

Figure 1. Mean variance in repeat number determined for 50 evenly spaced microsatellites over 200 simulation runs. For each of the simulations a selective sweep was assumed to have occurred at the microsatellite No. 25, which shows the most pronounced reduction in variability. Computer simulations were performed with a computer program written by Y. Kim and modified for microsatellites by T. Wiehe. Simulation parameters were: microsatellite spacing = 12 kb, tau = 0.001, s = 0.002, $\theta = 5$, $r = 5 \times 10^{-9}$.

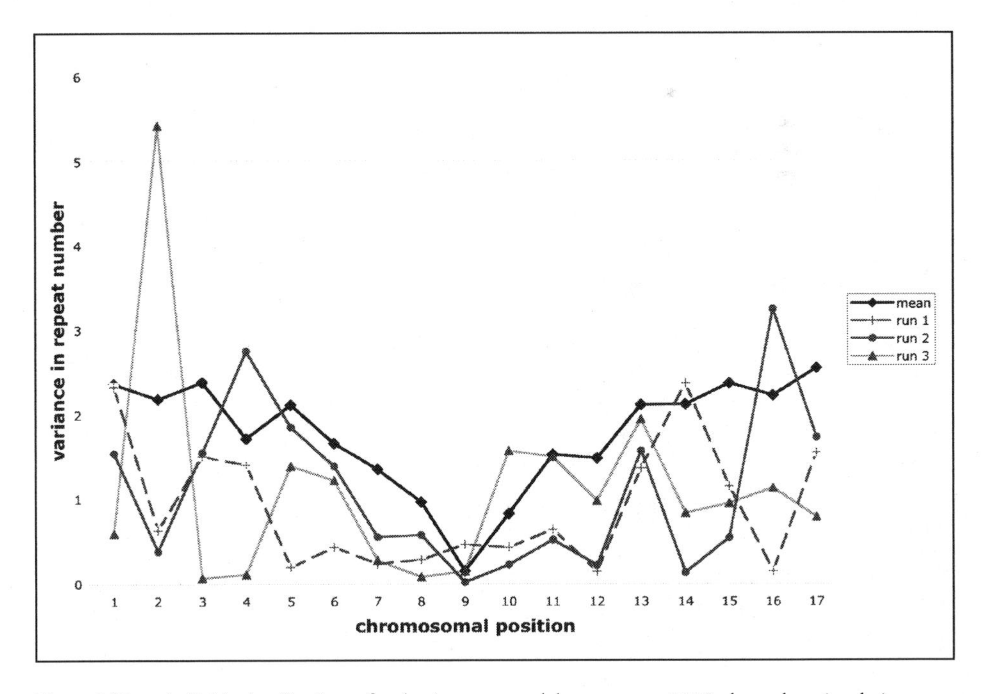

Figure 2. Three individual realizations of a selective sweep and the mean over 200 independent simulation runs.

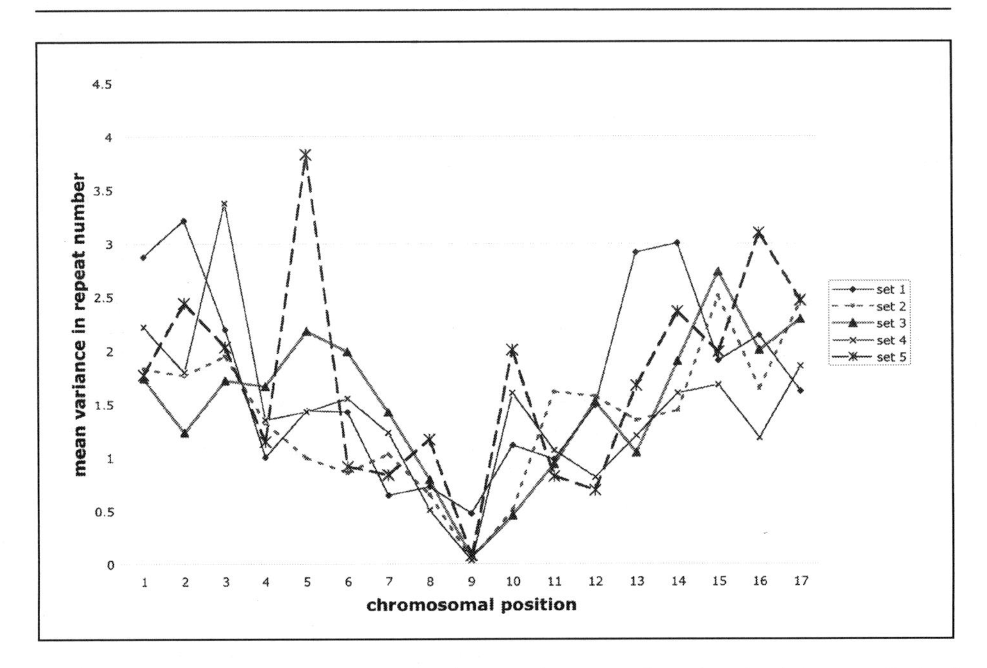

Figure 3. The mean variance in repeat number determined over 10 independent simulation runs. Each of the five sets includes data from 10 independent simulations.

example for selective forces operating on geographically isolated populations. Indeed, for allozymes, several parallel patterns of clinal variation have been described in *Drosophila melanogaster.*[29-32] Also selection for milk and meat production traits in cattle may provide an excellent example for parallel selection in genetically distinct populations. Nevertheless, the cases where such a mapping strategy could be applied are certainly limited. Furthermore, with epistasis different chromosomal regions may be selected in different populations, even though selective forces remain the same.

Alternatively, almost independent sweeps of the same beneficial mutation could be produced in populations connected by low levels of geneflow. Assuming that in one of the populations a beneficial mutation arises and sweeps through the population, this creates one selective sweep. For low rates of geneflow the beneficial mutation is exported to a neighboring population at a very low frequency. Thus, only a very small number of chromosomes with the beneficial mutation initiate a novel selective sweep in the next population. This could be regarded as a second realization of the spread of the same beneficial mutation. Naturally, the pattern of the sweep depends on several parameters, such as the rate of geneflow among the populations and the selection coefficient of the beneficial mutation. Furthermore, the mutation is assumed to have the same beneficial effect in all populations. Whether these conditions are met in natural populations still remains to be seen.

Identification of the Beneficial Mutation

For mutations shared among populations, it is possible that a mutation is beneficial in one population, but neutral or deleterious in another.[33] The above model of subsequent sweeps in populations connected by low levels of geneflow, however, assumes that a new mutation arises in one population first and subsequently migrates to others. In order to spread in other populations it is required that the mutation has the same/similar beneficial effect as in the original population. Assuming that the beneficial mutation has recently arisen, it is possible to identify the underlying molecular change by comparing sequences of the two groups of populations,

Box 1. Hitchhiking Mapping: Case Studies

For several species, *Drososphila melanogaster*, *D. simulans*, maize, and humans multiple, unlinked microsatellites have been analyzed and a number of candidate loci were identified.[21,24,34-37] Two important aspects emerged from these studies: first, the large number of loci analyzed results in a multiple testing problem; second, a large fraction of the loci surveyed were suspected to deviate from neutral expectations.

The analysis of multiple linked markers covering a genomic region that has been putatively subjected to directional selection could be used to distinguish between a selective and neutral scenario. While under neutrality alleles at linked markers are only weakly correlated (i.e., low linkage disequilibrium), after a recent selective sweep more linkage disequilibrium is expected (higher correlation). Nevertheless, a high density of markers may be required to detect the genomic region, which is affected by the selective sweep. *Plasmodium falciparum* offers a unique situation to study the consequences of a selective sweep. Natural populations have recently acquired resistance to antimalaria drugs and the molecular basis of the resistance is already understood. Furthermore, the microsatellite density in *Plasmodium falciparum* is extremely high, averaging one microsatellite/kb genomic sequence. Nair et al (2003) studied microsatellite variability around the dihydrofolate reductase gene, which is conferring resistance to the antimalaria drug pyrimethamine. As shown in Figure 4, a strong reduction in variability is observed around the target of selection (indicated by an arrow). The high density of microsatellites resulted in a correlated pattern at multiple loci flanking the target of selection. While the authors did not explore the range of neutral scenarios that could potentially also cause such a strong reduction at multiple flanking loci, it appears unlikely that such a pattern would be expected under neutrality. Formal test statistics based on the pattern of variability at flanking loci could be devised, which significantly reduce the rate of false positives compared to a single locus analysis (Wiehe, Nolte & Schlötterer, unpublished results).

The second problem, emerging from previous hitchhiking mapping studies, is that multiple loci may have been subjected to directional selection. Interestingly, for several species, which originated in Africa and colonized the rest of the world only recently, a higher number of selective sweeps has been proposed for the X-chromosome of nonAfrican populations.[24,35,37-39] In *D. melanogaster* and *D. simulans*, approximately 10% of the analyzed loci deviate from neutral expectations.[24,37,39] Thus, if the full X-chromosomal data set is used for standardization in the lnRθ test, the power to detect loci affected by a selective sweep is dramatically reduced (as the mean is shifted towards negative values and the variance is increased). It has been proposed to use autosomal microsatellites as a reference data set for standardization.[24] The rationale is that the ratio of effective populations sizes in two populations is expected to be the same for autosomes and X-chromosomes. Nevertheless, if one of the sexes has a higher variance in reproductive success (e.g., a small number of males mate with all females), the effective population size of X-chromosomes and autosomes deviates from the expected 3:4 ratio. Also population size changes have a different effect on both sets of chromosomes. Schöfl & Schlötterer (2004) used another approach for the identification of loci deviating from neutral expectations. Given that under neutrality lnRθ values are normally distributed, they iteratively removed outliers until no improved fit to a Gaussian distribution could be achieved. While the removal of 7 outliers improved the fit to a normal distribution for the X-chromosomal data set, not a single locus could be removed for the autosomal data set.

one with—and one without—a selective sweep. The beneficial mutation is expected to be absent (or at a low frequency) in the population without the selective sweep, but should be fixed or at a high frequency in populations showing the signature of a selective sweep. Hence, selective sweeps that occurred in several populations will not only provide a more accurate position of the target of selection, but they will also make the identification of the selected site easier, as this mutation is expected to be consistently at a high frequency in all populations showing the signature of a selective sweep.

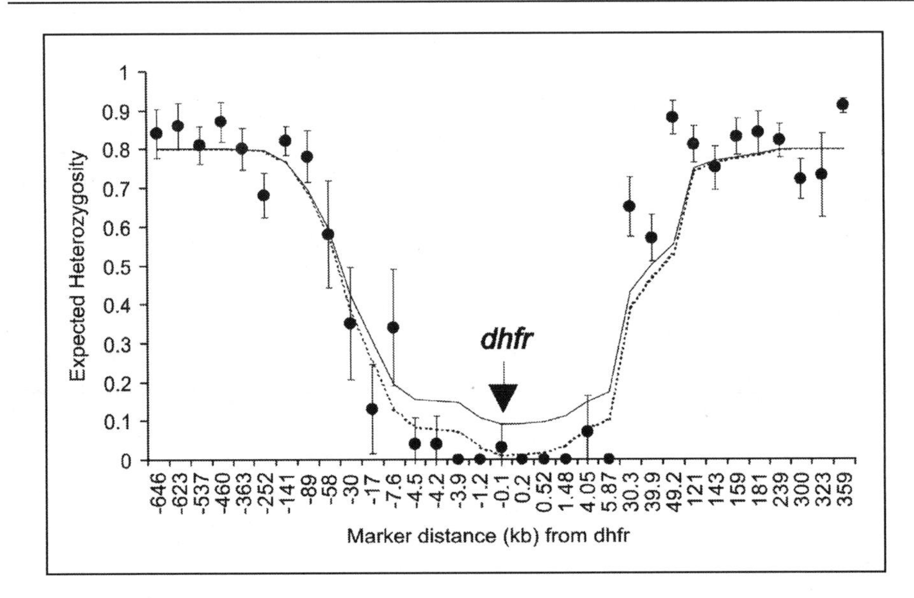

Figure 4. Pattern of microsatellite variability around *dhfr* in *Plasmodium falciparum*. The black and dotted lines show the expected heterozygosity (based on the deterministic hitchhiking model of[27] using different mutation parameters.[40] Figure reproduced from reference 40, with kind permission of Oxford University Press.

Acknowledgements

This work has been supported by Fonds zu Förderung der Wissenschaftlichen Forschung grants to CS. Many thanks to Y. Kim and T. Wiehe for sharing computer simulation code. B. Harr, J.-M. Gibert and C. Vogl provided helpful comments on the manuscript.

References

1. Feder ME, Mitchell-Olds T. Evolutionary and ecological functional genomics. Nat Rev Genet 2003; 4(8):651-657.
2. Schlötterer C. Towards a molecular characterization of adaptation in local populations. Current Opinions in Genetics & Development 2002; 12:683-687.
3. Schlötterer C. Hitchhiking mapping-functional genomics from the population genetics perspective. Trends Genet 2003; 19(1):32-38.
4. Harr B, Kauer M, Schlötterer C. Hitchhiking mapping-a population based fine mapping strategy for adaptive mutations in D. melanogaster PNAS 2002; 99:12949-12954.
5. Kohn MH, Pelz HJ, Wayne RK. Natural selection mapping of the warfarin-resistance gene. PNAS 2000; 97(14):7911-7915.
6. Maynard Smith J, Haigh J. The hitch-hiking effect of a favorable gene. Genet Res 1974; 23:23-35.
7. Otto SP. Detecting the form of selection from DNA sequence data. TIG 2000; 16:526-529.
8. Kreitman M. Methods to detect selection in populations with applications to the human. Annu Rev Genomics Hum Genet 2000; 1:539-559.
9. Tajima F. Statistical method for testing the neutral mutation hypothesis by DNA polymorphism. Genetics 1989; 123:585-595.
10. Nielsen R. Statistical tests of selective neutrality in the age of genomics. Heredity 2001; 86:641-647.
11. Luikart G, England PR, Tallmon D et al. The power and promise of population genomics: From genotyping to genome typing. Nat Rev Genet 2003; 4(12):981-994.
12. Litt M, Luty JA. A hypervariable microsatellite revealed by in vitro amplification of a dinucleotide repeat within the cardiac muscle actin gene. Am J Hum Genet 1989; 44:397-401.
13. Weber JL, May PE. Abundant class of human DNA polymorphisms which can be typed using the polymerase chain reaction. Am J Hum Genet 1989; 44:388-396.
14. Tautz D. Hypervariability of simple sequences as a general source for polymorphic DNA markers. NAR 1989; 17:6463-6471.

15. Schlötterer C. The evolution of molecular markers-just a matter of fashion? Nat Review Genet 2004; 5:63-69.
16. Schlötterer C. Evolutionary dynamics of microsatellite DNA. Chromosoma 2000; 109:365-371.
17. Ellegren H. Microsatellite mutations in the germline: Implications for evolutionary inference. Trends Genet 2000; 16(12):551-558.
18. Ohta T, Kimura M. A model of mutation appropriate to estimate the number of electrophoretically detectable alleles in a finite population. Genet Res 1973; 22:201-204.
19. Di Rienzo A, Donnelly P, Toomajian C et al. Heterogeneity of microsatellite mutations within and between loci and implications for human demographic histories. Genetics 1998; 148:1269-1284.
20. Harr B, Zangerl B, Brem G et al. Conservation of locus specific microsatellite variability across species: A comparison of two Drosophila sibling species D. melanogaster and D simulans MBE 1998; 15:176-184.
21. Schlötterer C. A microsatellite-based multilocus screen for the identification of local selective sweeps. Genetics 2002; 160(2):753-763.
22. Schlötterer C, Dieringer D. A novel test statistic for the identification of local selective sweeps based on microsatellite gene diversity. In: Nurminsky DI, ed. Selective Sweep. Georgetown: Landes Bioscience, 2005:55-64.
23. Moran PAP. Wandering distributions and electrophoretic profile. Theoretical Population Biology 1975; 8:318-330.
24. Kauer MO, Dieringer D, Schlötterer C. A microsatellite variability screen for positive selection associated with the "out of Africa" habitat expansion of Drosophila melanogaster. Genetics 2003; 165(3):1137-1148.
25. Schlötterer C, Kauer M, Dieringer D. Allele excess at neutrally evolving microsatellites and the implications for tests of neutrality. Proc Roy Soc Lond B 2004; in press.
26. Schlötterer C, Wiehe T. Microsatellites, a neutral marker to infer selective sweeps. In: Goldstein D, Schlötterer C, eds. Microsatellites-Evolution and Applications. Oxford: Oxford University Press, 1999:238-248.
27. Wiehe T. The effect of selective sweeps on the variance of the allele distribution of a linked multi-allele locus-hitchhiking of microsatellites. Theoretical Population Biology 1998; 53:272-283.
28. Kim Y, Stephan W. Detecting a local signature of genetic hitchhiking along a recombining chromosome. Genetics 2002; 160(2):765-777.
29. Singh RS, Long A. Geographic variation in Drosopila: From molecules to morphology and back. TREE 1992; 7(10):340-345.
30. van Delden W, Kamping A. Worldwide latitudinal clines for the alcohol dehydrogenase polymorphism in Drosophila melanogaster: What is the unit of selection. EXS 1997; 83:97-115.
31. Gilchrist AS, Partridge L. A comparison of the genetic basis of wing size divergence in three parallel body size clines of Drosophila melanogaster. Genetics 1999; 153(4):1775-1787.
32. Calboli FC, Kennington WJ, Partridge L. QTL mapping reveals a striking coincidence in the positions of genomic regions associated with adaptive variation in body size in parallel clines of Drosophila melanogaster on different continents. Evolution 2003; 57(11):2653-2658.
33. Orr HA, Betancourt AJ. Haldane's sieve and adaptation from the standing genetic variation. Genetics 2001; 157(2):875-884.
34. Vigouroux Y, McMullen M, Hittinger CT et al. Identifying genes of agronomic importance in maize by screening microsatellites for evidence of selection during domestication. PNAS 2002; 99(15):9650-9655.
35. Payseur BA, Cutter AD, Nachman MW. Searching for evidence of positive selection in the human genome using patterns of microsatellite variability. Mol Biol Evol 2002; 19(7):1143-1153.
36. Kayser M, Brauer S, Stoneking M. A genome scan to detect candidate regions influenced by local natural selection in human populations. Mol Biol Evol 2003; 20(6):893-900.
37. Schöfl G, Schlötterer C. Patterns of microsatellite variability among X chromosomes and autosomes indicate a high frequency of beneficial mutations in nonAfrican D. simulans. Mol Biol Evol 2004.
38. Kauer M, Zangerl B, Dieringer D et al. Chromosomal patterns of microsatellite variability contrast sharply in African and nonAfrican populations of Drosophila melanogaster. Genetics 2002; 160:247-256.
39. Glinka S, Ometto L, Mousset S et al. Demography and natural selection have shaped genetic variation in Drosophila melanogaster: A multi-locus approach. Genetics 2003; 165(3):1269-1278.
40. Nair S, Williams JT, Brockman A et al. A selective sweep driven by pyrimethamine treatment in southeast asian malaria parasites. Mol Biol Evol 2003; 20(9):1526-1536.

Chapter 10

Understanding the Functional Importance of Human Single Nucleotide Polymorphisms

Saurabh Asthana and Shamil Sunyaev*

Abstract

Single nucleotide polymorphisms (SNPs) are the major source of human genetic variation, and the functional subset of SNPs, predominantly in protein coding regions, contributes to phenotypic variation. However, much of the variation in coding regions may not produce any functional effects. There are two broad strategies for classifying polymorphism as functional or neutral: sequence-based methods predict functional significance based on conservation scores calculated from alignments of homologous gene sequences; structure-based methods map variations to known protein structures and predict likely effects based on properties of proteins. Several tools have been developed to classify polymorphism as functional or neutral based on these methods. It was shown that most of functional SNPs are evolutionarily deleterious. Though the utility of the tools has not yet been adequately demonstrated, they may have important applications in the area of medical genetics.

Introduction

The observation that organisms differ from each other extends back to the earliest human history. Aristotle developed taxonomic systems to categorize diverse populations into hierarchies, recognizing that the degree of differentiation between organisms corresponds to the degree of their separation by familial relationships. By now we have come to understand that inherited differences are transmitted via genetic material, and that the differences between individuals must ultimately translate into differences in their genetic sequence. One of the enduring puzzles of biology is understanding variation—what is it that makes sister different from sister? How do these changes in genetic material manifest as changes in outward appearance, behavior or biochemical makeup?

At its most basic level, genetic variation consists of simple changes in sequence—base-pair substitutions, insertions and deletions. What is commonly understood by the term "allele", i.e., two functionally divergent forms of the same gene, in the end might consist of only a single differing nucleotide base-pair. The vast majority (90%) of genetic variation in humans consists of single nucleotide polymorphisms (SNPs).[1] But all of this variation need not translate into observable phenotypic variation; most of it will be functionally neutral. The majority of SNPs occur in intergenic or intronic noncoding regions of the genome. Most noncoding SNPs are unlikely to have a functional impact; only a small minority is believed to have

*Corresponding Author: Shamil Sunyaev—Genetics Division, Department of Medicine, Brigham and Women's Hospital and Harvard Medical School, Harvard Medical School New Research Building, 77 Ave. Pasteur, Boston, Massachusetts 02115, U.S.A . Email: ssunyaev@rics.bwh.harvard.edu.

Discovering Biomolecular Mechanisms with Computational Biology, edited by Frank Eisenhaber. ©2006 Landes Bioscience and Springer Science+Business Media.

functional significance, predominantly due to the effect on gene expression. Of the fraction of SNPs that does occur in coding sequence approximately half are synonymous substitutions, which rarely produce an observable effect on phenotype. Even the remainder, which is guaranteed to result in amino acid variations, does not necessitate functionally divergent protein products.

Although some variants in noncoding regions could have phenotypic effects, the great majority of functional variation likely falls in coding regions. The structure of intergenic regions are also so poorly understood that they are largely impenetrable to analysis at the moment. For this reason, we limit our consideration of variation to nonsynonymous coding SNPs.

Identifying functional variation might be valuable in several contexts. First, we would expect the majority of functional variation to be detrimental (since beneficial variation is believed to be rare). Specifying functionally significant sequence divergence could therefore provide avenues to understanding disease susceptibility. Second, pinpointing the functional significance of nucleotide substitutions between species may shed light on the basic mechanisms of evolution, and reveal how genetic variation is translated into phenotypic variation.

There are a number of strategies we may follow to answer this question.

Comparative Sequence Analysis

Because purifying selection will eliminate variation at functionally important positions, as genes evolve and diverge functionally important positions will show greater conservation between species. Since selection operates exactly on the basis of phenotypic significance, conservation should be expected to be an excellent guide to functionality.

The availability of sequence information from hundreds of species allows the quick retrieval of many protein homologs of a gene of interest. A number of standard techniques exist for constructing multiple alignments of homologous sequences.

A very rude measure of conservation can be obtained simply by examining the degree of entropy at a particular position in a multiple sequence alignment of homologous protein sequences. Low entropy, i.e., high conservation, would suggest the position is important. This simple measure has been shown to be an effective discriminator for functionality.[2]

A more sophisticated measure based on multiple sequence alignments uses position-specific scoring matrices. A probability measure of the likelihood that a variant is permissible (profile score) may be generated for each amino acid at each position. This profile score accounts for phylogenetic history and amino acid frequency as well as conservation. Profile scores are employed in two tools that predict functional variation based on multiple sequence alignments, PolyPhen[3] and SIFT.[4]

Any set of homologous sequences is presumably descended from a common ancestral sequence. Accordingly, common sequence identity may be the result of common descent, making it necessary to segregate the effects of phylogeny from the effects of selection on conservation.

Most profile scores employ sequence-weighting techniques to discriminate between the effect of selection and simple phylogenetic proximity, so that closely related sequences are downweighted. In the ideal each sequence is weighted according to the information it adds to the alignment with regard to the effect of selection. Sequences that are phylogenetically more distant from the others in the alignment will provide the most information—if an amino acid has been conserved across a huge evolutionary distance it is highly likely to be important. Two basic strategies exist for determining evolutionary distance. One is to weight the sequences according to a reconstruction of their phylogenetic tree. The other is to weight sequences according to some metric based on sequence divergence, e.g., pairwise identity or from the degree of identity at aligned positions.[5] Most sequence-weighting techniques apply the same weight to the entire sequence, but some give position-specific weight.

A separate strategy for quantifying conservation is also based on phylogeny, counting the minimum number of amino acid substitutions from an ancestral sequence required to produce the pattern of variation in a multiple sequence alignment.[7]

However, this strategy does not include any information about amino acid frequencies; the pattern of variation at a locus can also provide meaningful information. For example, if an amino acid occurs frequently at a particular position in an alignment it is unlikely to have a deleterious effect; a rare variant is still likely to be tolerated. Amino acids that do not occur in the alignment are much more likely to have deleterious effects. An alternative, broader measure is to compare amino acids by class (e.g., if no hydrophobic amino acids occur in the alignment, a hydrophobic variant is likely to be deleterious). This property has been shown to be quite an effective classifier of variation on its own.[8]

Multiple sequence alignments may also suffer from a paucity of informative sequences. Consider the case of a sequence with a meager two homologs; in this instance conservation may be misapprehended simply because there are not enough sequences to allow the full range of variation. This lack may be retrieved by adding Bayesian pseudocounts to profile scores based on background levels of variation.[4]

Conservation becomes more significant as evolutionary time of divergence increases. However, conservation (or lack of it) is only meaningful to the extent that function is preserved. For example, HIV-1 protease aligns well against the proteases from Rous Sarcoma Virus and Avian Myeloblastosis Virus; however, because the substrate specificity of these proteins has changed relative to HIV-1 protease, excluding these two sequences from the alignment can improve the predictive ability of the method. Since there is no simple way to recognize functional divergence from sequence, it is necessary to inspect alignments by hand and cull problematic sequences.[4]

Additionally sequence comparison-based methods will not deal well with compensatory substitutions. A certain variation in one phylogenetic line might be tolerated simply because there is a compensatory mutation at another site; however, the same variant in a different phylogenetic line that lacks the compensatory mutation might prove deleterious. However, since the variant is apparently "tolerated" according to the pattern of variation at that position, the above methods will judge it to be benign.[3]

Structure-Based Methods

Ultimately a gene expresses its function through its protein product. Examination of the three-dimensional structure of a protein can reveal aspects of a particular variation that are not readily evident from the sequence itself. Many of the insights gained from the study of protein structure can be applied to this problem.

Structural information can help identify functional sites and sites in proximity to functionally important regions. However, systematic studies demonstrate that the great majority of functional mutations affect protein stability.[9] As accurate computational methods for predicting the effect of amino acid substitutions on protein stability are not available yet, a number of empiric parameters related to protein stability have been proposed as predictors. These included accessible surface area coupled with hydrophobic propensities, packing parameters, hydrogen bonding, conformational propensities, crystallographic B-factors and others. All parameters were shown to correlate with the effect of substitutions. Many of the parameters did not result in improvement of the prediction because of an unacceptable rate of false-positives,[10] whereas a few were shown to produce a small number of false-positives and thus appear to be useful predictors.

However, in general structural classification is notoriously weak compared to sequence-conservation based classification. For this reason few methods rely solely on structural methods; many more combine both types of analysis.

Combined Methods

A number of predictive methods combine structure and sequence-based methods, including the PolyPhen tool[3] and the MutDB tool.[11]

Because information derived from structural properties of genes is wholly independent of information derived from sequence conservation, the combination can be vastly improved in performance. In addition the failings of either method may be orthogonal; e.g., structurebased methods will not misjudge a compensatory mutation as comparative-sequence-based methods will (discussed above).

Systematic analysis by Saunders and Baker[12] confirms that errors in the functional classification of variants can be significantly reduced by combining structural and comparative sequence information. They assess the performance of the SIFT tool when augmented with various combinations of structural rules and find strong improvements in classification errors when compared to the performance of SIFT alone. The most effective combination according to their analysis included two structural features: the density of c^{β} atoms around a residue (a measure of how deeply buried the site is), and a normalized *B*-factor (a measure of side-chain flexibility at a site).

Methods of Validation

Two validation strategies have been proposed for testing the methods for predicting the functional significance of polymorphic variants. The first approach utilizes data on genetic variants that are already known to be functionally significant. There are several compendiums of such information, including the Online Mendelian Inheritance in Man (OMIM) database (http://www.ncbi.nlm.nih.gov/omim/), which catalogues human disease alleles; the Human Gene Mutation Database (HGMD),[1] a catalogue of mutations in human nuclear genes causing disease; and the Human Genome Variation Database (HGVBase),[14] which is a general catalogue of genotype/phenotype associations.

Disease mutations will only encompass a specific range of functional variation, since only fully penetrant genes with Mendelian inheritance will produce disease phenotypes. Such datasets can therefore not be assumed to be free of bias towards particular mutations.

As a result of validation experiments on this dataset, the PolyPhen tool, for example, was shown to correctly predict 82% of disease mutations (57% under a more stringent threshold). The test was performed on proteins with available multiple homologs and 3D-structure, therefore the accuracy of the method might be slightly lower when applied to all human amino acid polymorphisms.

To estimate the rate of false-positive predictions, substitutions between human proteins and closely related mammalian orthologues are examined as putatively neutral variants. PolyPhen produces 8% false-positives on this dataset (3% under a more stringent threshold).

A second, complementary validation strategy relies on site-directed mutagenesis experiments.[12] Systematic in vitro mutation screens can identify specific point mutations that disrupt the function of model proteins. Since these screens can cover the full sequence, they are often taken to be devoid of bias towards any particular class of mutations; they will encompass severe mutations as well as less functional ones. However, since such screens rely on gross biochemical effects and do not place the mutation in its full context, they will miss subtler effects on phenotype that will be covered by disease mutations. Additionally such screens rely on a small number of proteins that may not be representative of all possible protein structures, and may not come from the organism of interest. If the model protein is biased towards particular features (e.g., a hydrophobic interface), the importance of certain functional classes of mutations might be overrepresented.

These data sources may also be used in combination with machine-learning methods as training sets to select the best classifiers from a wide variety of structural and sequence comparative metrics.[8]

Analysis of natural selection acting on deleterious variants (discussed below) can be also used as an independent validation of the prediction techniques.

Detecting the Effects of Selection

The functional significance of variation is precisely the characteristic that selection operates; thus, identifying exactly where selection has acted should be an exceptional method of determining significance. Unfortunately, it is extraordinarily difficult to determine whether selection has acted on specific positions. For example, the effects of population history are often indistinguishable from the effects of selection. However, identifying the effects of selection can be used to validate putative functional SNPs. Additionally genome-wide studies of selection can provide quantitative assessment of the proportion of SNPs expected to be important.

That the genome is under selection is immediately obvious. A straightforward comparison of amino-acid polymorphism rate to synonymous polymorphism rate reveals that amino-acid polymorphisms occur at 1/3-1/5 the frequency of synonymous polymorphism, suggesting that 70-80% of amino-acid polymorphisms are deleterious.[15] Comparison of variation in allele frequency between human subpopulations demonstrates that the allele frequency spectrum is shifted towards rare (low and high frequency) variants, indicating the influence of selection.[16]

The distribution of allele frequency for polymorphic loci further demonstrates the effect of selection on the genome. A comparison of the frequency of amino-acid polymorphism to synonymous (presumably neutral) polymorphism shows that the mean frequency of amino-acid polymorphism is significantly lower, suggesting that many of these loci are experiencing purifying selection.[17] Analysis of the distribution of polymorphic variants among major human subpopulations using the F_{ST} statistic also showed that many genes are under selective pressure.[18]

Comparison of polymorphism with species divergence and analysis of allele frequency distribution suggests that nonsynonymous SNPs predicted to be functional by methods described above are evolutionarily deleterious, i.e., most of them are under negative selective pressure.[19] This observation logically links effects on protein function and stability with effects on evolutionary fitness.

Genome-Wide Analysis of Functional Polymorphic Variants

The primary ingredient in identifying functionally important variation is a comprehensive catalogue of genetic variation in general. Human variation, though not exhaustively enumerated, has been significantly covered; at present some 4.3 million SNPs have been validated. This data is available in the dbSNP database (http://www.ncbi.nlm.nih.gov/SNP/). The bulk of SNP data comes from two sources. About half is taken from overlapping sequence produced in the course of the human genome sequencing project; accordingly, these SNPs are not randomly distributed in the genome and are therefore sometimes excluded from experiments. The bulk of remaining SNPs come from The SNP Consortium (TSC),[20] which attempts to exhaustively enumerate SNPs by resequencing the genomes of 24 unrelated individuals. TSC SNPs are randomly distributed and also often have associated allele frequency information available. SNPs have been contributed by many other sources as well.

Methods for predicting functional effect of nonsynonymous SNPs and methods to detect natural selection were both applied to SNPs from public databases. Though these SNPs are taken from generally healthy individuals, it was shown that individual human genomes have a large number of potentially deleterious polymorphic variants. Estimates of the standing number of deleterious mutations in an individual genome mostly reach a consensus of ~500-2000[21,16] mutations, although a much higher estimate of ~6000-13,000 mutations was reported[8] (though the latter is probably an overestimate, since it does not account for the depressed allele frequencies of deleterious mutations).

Significance for Medical Genetics

Understanding functional variation is a potential goldmine in the study of inherited disease. Many genetic diseases with simple Mendelian patterns of inheritance are well understood, but increasingly research efforts are running up against complex diseases. Though there has been no demonstration of the utility of the above predictive methods, they may prove useful in elucidating the genetic basis of complex diseases.

Two different hypotheses exist regarding the nature of the genetic basis of common human disease phenotypes.

The CD/CV (common disease - common variant)[22,23] hypothesis postulates that most disease susceptibility variants are common, and in the majority of cases, a limited number of alleles in a limited number of loci account for the majority of disease susceptibility. In the extreme case, even a single common allelic variant in a single locus may be responsible for a common disease phenotype.[24]

A possible evolutionary basis for the CD/CV hypothesis is provided by the trade-off (antagonistic pleiotropy) model.[25] This model suggests that deleterious disease susceptibility alleles can spread in the population because of a favorable effect on another trait. Since many human common diseases are late-onset phenotypes, the pressure of negative selection against them is reduced, and it is hypothesized that the same allele may have a positive effect at an early age.

The alternative, the CD/RV (common disease - rare variant) hypothesis, states that most of the common human diseases are due to a large number of possibly rare variants in many loci.[26] The CD/RV hypothesis can most easily be explained by a mutation accumulation model, which assumes that the genetic basis of a common disease can be explained by the accumulation of many allelic variants, each under low selective pressure. The accumulation of these variants, with some of them reaching significant frequencies, is possible for two reasons. First, the intensity of selection may be low, especially for late-onset diseases. Second, multiple variants in multiple loci contribute to disease susceptibility, and thus can be kept in a mutation-selection balance.

The accumulation of slightly deleterious variants in multiple interacting loci produces a complex picture of common disease susceptibility inheritance under the CD/RV hypothesis. Differences between the CD/CV and CD/RV hypotheses are critical to the design of studies aimed at the identification of disease susceptibility alleles.

The studies described in previous sections show that computational methods are able to predict and quantify deleterious SNP alleles kept under mutation- selection balance, which are important in the inheritance of common disease under the mutation accumulation model. The observation of a large number of these alleles may be considered as indirectly supporting the mutation accumulation model. Obviously, effects on molecular function are an essential feature of susceptibility alleles under both models.

Linkage studies have proven very successful in mapping the loci responsible for Mendelian diseases, but are much less efficient when studying complex phenotypes with non-Mendelian patterns of inheritance.[27] Association studies that test allele frequencies in disease populations versus healthy controls are hypothesized to be better suited for complex phenotype analysis.[28] Under the CD/CV hypothesis, whole-genome association studies might prove useful. These studies can be based on linkage disequilibrium between marker SNPs and causative alleles in the same haplotype blocks.[29] Possible allelic and nonallelic heterogeneity and the presence of epistatic interactions are the major complications of this approach.

An alternative approach, which might work even in complex situations, is based on candidate genes or candidate genomic regions. In addition to the careful selection of candidate genes (or loci) in this approach, proper prioritization of SNPs for the analysis is essential. This prioritization might be achieved via computational methods.

Testing SNP alleles on a genome-wide scale requires a massive amount of genotyping and, most importantly, strongly reduces the statistical efficiency of association studies because of multiple test corrections. Focusing on potentially functional SNPs first will help to overcome both of these problems. Conversely, a group of SNPs in linkage disequilibrium might be found to be statistically associated with a phenotype of interest, but identification of the causative variant among them may present a challenge. Computational methods could be used to predict the functional variant.

Under the CD/RV hypothesis, association studies testing for the frequency of individual alleles might fail in many cases.[30-32] However, if a good set of multiple candidate genes is available with a reasonable degree of confidence by using methods for predicting the functional effect of SNPs, it is possible to test whether the number of alleles predicted to be deleterious is higher in all of these genes in patients as opposed to controls. This might provide a direct test for accumulation of deleterious mutations responsible for a specific phenotype.

References

1. Collins F, Brooks L, Chakravarti A. A dna polymorphism discovery resource for research on human genetic variation. Genome Res 1998; 8:703-705.
2. Sunyaev S, Ramensky V, Bork P. Towards a structural basis of human nonsynonymous single nucleotide polymorphisms. Trends Genet 2000; 16:198-200.
3. Ramensky V, Bork P, Sunyaev S. Human nonsynonymous snps: Server and survey. Nucleic Acids Res 2002; 30:3894-3900.
4. Ng P, Henikoff S. Predicting deleterious amino acid substitutions. Genome Res 2001; 11:863-874.
5. Henikoff S, Henikoff J. Position-based sequence weights. J Mol Biol 1994; 243:574-578.
6. Sunyaev S, Eisenhaber F, Rodchenkov I et al. Psic: Profile extraction from sequence alignments with position-specific counts of independent observations. Protein Eng 1999; 12:387-394.
7. Miller M, Kumar S. Understanding human disease mutations through the use of interspecific genetic variation. Hum Mol Genet 2001; 10:2319-2328.
8. Chasman D, Adams R. Predicting the functional consequences of nonsynonymous single nucleotide polymorphisms: Structurebased assessment of amino-acid variation. J Mol Biol 2001; 307:683-706.
9. Wang Z, Moult J. Snps, protein structure and disease. Hum Mutat 2001; 17:263-270.
10. Ferrer-Costa C, Orozco M, de la Cruz X. Characterization of disease-associated single amino acid polymorphisms in terms of sequence and structure properties. J Mol Biol 2002; 315:771-786.
11. Mooney S, Altman R. Mutdb: Annotating human variation with functionally relevant data. Bioinformatics 2003; 19:1858-1860.
12. Saunders C, Baker D. Evaluation of structural and evolutionary contribution to deleterious mutation prediction. J Mol Biol 2002; 322:891-901.
13. Krawczak M, Cooper D. The human gene mutation database. Trends Genet 1997; 13:121-122.
14. Fredman D, Siegfried M, Yuan Y et al. Hgvbase: A human sequence variation database emphasizing data quality and a broad spectrum of data sources. Nucleic Acids Res 2002; 30:387-391.
15. Halushka M, Fan J, Bentley K et al. Patterns of single-nucleotide polymorphisms in candidate genes for blood-pressure homeostasis. Nat Genet 1999; 22:239-247.
16. Fay J, Wyckoff G, Wu C. Positive and negative selection on the human genome. Genetics 2001; 158:1227-1234.
17. Hughes A, Packer B, Welch R et al. Widespread purifying selection at polymorphic sites in huam protein-coding loci. Proc Natl Acad Sci USA 2003; 100:15754-15757.
18. Akey J, Zhang G, Zhang K et al. Interrogating a high-density snp map for signatures of natural selection. Genome Res 2002; 12:1805-1814.
19. Sunyaev S, Kondrashov F, Bork P et al. Impact of selection, mutation rate and genetic drift on human genetic variation. Hum Mol Genet 2003; 12:3325-3330.
20. Group TISMW. A map of human genome sequence variation containing 1.4 million snps. Nature 2001; 409:928-933.
21. Sunyaev S, Ramensky V, Kock I et al. Prediction of deleterious human alleles. Hum Mol Genet 2001; 10.
22. Lander E. The new genomics: global views of biology. Science 1996; 274:536-539.
23. Chakravarti A. Population genetics—making sense out of sequence. Nat Genet 1999; 21:56-60.
24. Reich D, Lander E. On the allelic spectrum of human disease. Trends Genet 2001; 17:502-510.
25. Partridge L, Gems D. Mechanisms of ageing: Public or private? Nat Rev Genet 2002; 3:165-175.
26. Wright A, Charlesworth B, Rudan I et al. A polygenic basis for late-onset disease. Trends Genet 2003; 19:97-106.
27. Altmüller J, Palmer L, Fischer G et al. Genomewide scans of complex human diseases: True linkage is hard to find. Am J Hum Genet 2001; 69:936-950.
28. Risch N, Merikangas K. The future of genetic studies of complex human diseases. Science 1996; 273:1516-1517.
29. Gabriel S, Schaffner S, Nguyen H et al. The structure of haplotype blocks in the human genome. Science 2002; 296:2225-2229.
30. Wright A, Carothers A, Pirastu M. Population choice in mapping genes for complex diseases. Nat Genet 1999; 23:397-404.
31. Pritchard J, Cox N. The allelic architecture of human disease genes: Common disease- common variant. or not? Hum Mol Genet 2002; 11:2417-2423.
32. Terwilliger J, Haghighi F, Hiekkalinna T et al. A bias-ed assessment of the use of snps in human complex traits. Curr Opin Genet Dev 2002; 12:726-734.

Chapter 11

Correlations between Quantitative Measures of Genome Evolution, Expression and Function

Yuri I. Wolf, Liran Carmel and Eugene V. Koonin*

Abstract

In addition to multiple, complete genome sequences, genome-wide data on biological properties of genes, such as knockout effect, expression levels, protein-protein interactions, and others, are rapidly accumulating. Numerous attempts were made by many groups to examine connections between these properties and quantitative measures of gene evolution. The questions addressed pertain to the most fundamental aspects of biology: what determines the effect of the knockout of a given gene on the phenotype (in particular, is it essential or not) and the rate of a gene's evolution and how are the phenotypic properties and evolution connected? Many significant correlations were detected, e.g., positive correlation between the tendency of a gene to be lost during evolution and sequence evolution rate, and negative correlations between each of the above measures of evolutionary variability and expression level or the phenotypic effect of gene knockout. However, most of these correlations are relatively weak and explain a small fraction of the variation present in the data. We propose that the majority of the relationships between the phenotypic ("input") and evolutionary ("output") variables can be described with a single, composite variable, the gene's "social status in the genomic community", which reflects the biological role of the gene and its mode of evolution. "High-status" genes, involved in house-keeping processes, are more likely to be higher and broader expressed, to have more interaction partners, and to produce lethal or severely impaired knockout mutants. These genes also tend to evolve slower and are less prone to gene loss across various taxonomic groups. "Low-status" genes are expected to be weakly expressed, have fewer interaction partners, and exhibit narrower (and less coherent) phyletic distribution. On average, these genes evolve faster and are more often lost during evolution than high-status genes. The "gene status" notion may serve as a generator of null hypotheses regarding the connections between phenotypic and evolutionary parameters associated with genes. Any deviation from the expected pattern calls for attention—to the quality of the data, the nature of the analyzed relationship, or both.

Introduction

Quantitative genomics involves numerous measures reflecting different aspects of the evolutionary history and the physiological role of a given gene (protein). One can estimate the evolution rate of a gene, measured in different organisms; its expression level in different tissues

*Corresponding Author: Eugene V. Koonin—National Center for Biotechnology Information, National Library of Medicine, National Institutes of Health, Bethesda, Maryland 20894, U.S.A. Email: koonin@ncbi.nlm.nih.gov.

Discovering Biomolecular Mechanisms with Computational Biology, edited by Frank Eisenhaber.

Table 1. Connections between various measures of sequence evolution rate, gene loss, expression, and fitness effect[a]

	K_{aa}	K_N	K_S	K_5	K_3	PGL	E_H	B_H	E_C	E_Y	E_Y
1 protein evolution rate (K_{aa})	x										
2 CDS non-synonymous evolution rate (K_N)	+	x									
3 CDS synonymous evolution rate (K_S)		+	X								
4 5'-UTR evolution rate (K_5)		+	+	x							
5 3'-UTR evolution rate (K_3)		+	+	+	x						
6 propensity for gene loss (PGL)	+					x					
7 expression level in human (E_H)	-	-	-	0	-	-	x				
8 expression breadth in human (B_H)	-	-	-	0	-		+	x			
9 expression level in *C. elegans* (E_C)	-					-	+		x		
10 expression level in *S. cereviseae* (E_Y)	-					-	+		+	x	
11 viability of gene disruption in *S. cereviseae* (E_Y)	+					+	-	-	-	-	x

[a] The data was from references15, 16.

and in different taxonomic groups; the tendency of a gene to be lost during evolution of different lineages of organisms or its tendency to produce paralogous copies via duplication; its position in the metabolic, signaling and protein interaction networks; and a variety of other quantities (e.g., refs. 1-4). Not unexpectedly, many of such measures are not independent. The literature on the subject (see specific references below) reports numerous positive and negative correlations: between the synonymous and nonsynonymous evolution rates within a gene; between evolution rate and expression level; between propensity of gene loss and fitness effect; and many more (Table 1). Some of these correlations are very strong for quite obvious reasons, such as evolution rates in different lineages or expression levels of orthologous genes; others are less trivial, e.g., the correlation between the degree of conservation of a gene's presence in different lineages and the degree of conservation of its sequence; yet others are remarkably low or absent, sometimes running contrary to expectations (evolution rate vs. number of protein-protein interactions or conservation of gene sequence and that of expression profiles).

Diverse as they are, all these purported correlations, except for the most obvious ones, share one somewhat disturbing feature: although they may be highly statistically significant due to the large number of data points, they typically explain only a small fraction of the variance of the analyzed quantities. Hence considerable debate around many of these observations, which is further compounded by problems with the completeness and quality of much of the data involved, particularly that coming from genome-scale analyses of gene expression, protein-protein interactions, and other aspects of gene functioning. For example, the argument about the link—or lack thereof—between the connectivity of a protein in protein-protein interaction networks and its evolutionary rate has already gone through at least three cycles of opposing claims, and there is still no definitive solution in sight.[5-10] Even when the existence of a link is not seriously questioned, as is, e.g., the case with the negative correlation between a gene's expression level and sequence evolution rate, the nagging question remains as to the ultimate importance of these observations. Given that the nontrivial correlations, however statistically significant they might be, are all relatively weak, it is quite legitimate and, probably, prudent to ask whether one should emphasize the existence of a particular link or the fact that the effect of one of the analyzed variables on the other(s) is only modest. Answering these questions is not easy, and yet, they are pressing because the higher-level problems addressed in this area of research are, arguably, among the most fundamental ones in biology, e.g., what determines the fitness effect of a gene's knockout or the rate of its evolution.

Quantitative genomics is a very young discipline which started in earnest only at the brink of the 21st century, when genome-wide data beyond the sequences themselves (gene expression, protein-protein interaction etc) began to accumulate. Nevertheless, in these few years, a fairly complex maze of observations on connections—or lack thereof—between all kinds of quantities has emerged. We believe that the field is in rather urgent need of a coherent conceptual framework that would allow one, simply put, to make sense of these diverse and often contradictory bits and pieces of information. Here, we present a brief overview of the available results on genomic correlations and discuss some preliminary glimpses of a would-be synthesis.

Evolution Rate, Expression Level and Expression Breadth

Numerous reports, including our own research, point to a significant correlation between the measures of evolutionary conservation of a protein and measures of its expression[11-16] (Table 1). Several notable conclusions emerge from these analyses. Firstly, there is a strong cohesion between measures of the same quantity obtained for different, in many cases, phylogenetically distant species. Despite obvious biological differences between, e.g., mammals, nematodes, and yeasts, expression levels of orthologous proteins from these species display positive correlations with r-values of 0.3-0.5 (with many hundreds of proteins in the dataset, the correlations are significant at p-values $<<10^{-10}$).[15] Likewise, evolution rates estimated for different lineages and across different ranges of distances tend to show even greater concordance (r-values of 0.7-0.9 between distantly related bacterial lineages).[17] Secondly, expression breadth, defined as the number of different tissues where a gene is significantly expressed, and the connectivity (node degree) in the gene coexpression network behave in essentially the same way as the expression level in expression *vs.* evolution rate comparisons. Specifically, there is a highly significant negative correlation between each of these parameters of gene expression and sequence evolution rate; in other words, highly and widely expressed genes, which have numerous coexpression partners, tend to evolve slowly (Fig. 1). Finally, while this negative correlation between evolution rate and expression parameters holds for the great majority of the relevant data, a notable exception breaks this nearly universal pattern. Analysis of mammalian microarray data shows that, while the synonymous and non-synonymous nucleotide evolution rates within the coding sequence and the nucleotide sequence evolution rate in the 3'-UTR behave as expected, the evolution rate of the 5'-UTR shows no correlation with expression.[16] This apparent discrepancy probably points to a distinct mode of evolution and/or a specific and still not understood connection between the sequence and its expression for the 5'-UTRs of (at least) mammalian genes (Fig. 1).

Evolution Rate, Gene Loss and Fitness Effect

The statement that sequence evolution rate and the tendency of a gene to be lost during evolution are correlated at first glance seems almost trivial—after all, it should be expected that evolutionarily conserved (i.e., slowly evolving) genes are also phylogenetically conserved, i.e., their orthologs more densely populate the tree of life than those of fast-evolving genes. In support of this straightforward line of reasoning, a highly significant correlation between

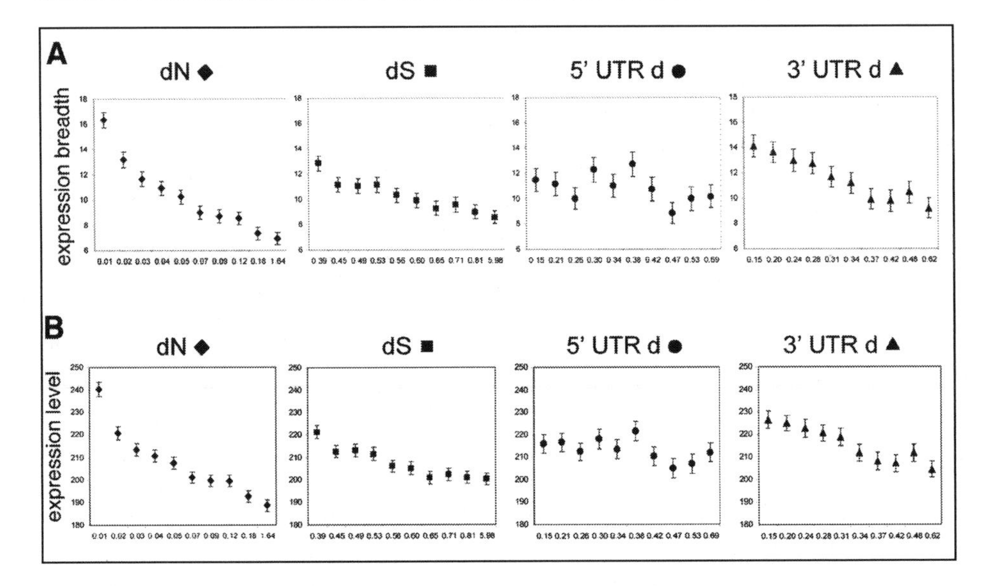

Figure 1. Correlation between evolution rate, expression breadth and expression level. Human microarray data from GEO database (GDS181.soft) were analyzed to determine expression breadth (a; number of tissues with expression level $\geq$200 AD) and expression level (b; the sum of the $\log_2$ normalized AD values over all tissues) as described previously.[16] Evolutionary distance between the human gene and its mouse ortholog was determined in nonsynonymous (dN) and synonymous (dS) sites of the coding region, 5'-UTRs (5'UTR d) and 3'-UTRs (3'UTR d). Genes were grouped according to the evolutionary distances (which, for orthologs, can be used as proxy for rates) in bins of approximately equal size; mean and variance of expression level and expression breadth were calculated for each bin.

these parameters has been observed[15] (Table 1, Fig. 2). However, these two faces of evolutionary conservation are not linked directly via a cause and effect relationship. Most likely, the strongest factor affecting the connection is the local (i.e., species-specific) fitness effect of the gene, usually measured as gene dispensability in knock-out experiments. It has been pointed out that genes experimentally shown to be essential tend to evolve slower than non-essential ones[18,19] although, again, the causal relationship between these parameters has been questioned.[20] Obviously, the gene dispensability over short evolutionary intervals entirely depends on the fitness effect of the gene loss (genes with a lethal knock-out phenotype in a particular species, by definition, cannot be lost in that species), while long-range loss propensity is more subtly determined by the evolution of the whole genome (changes in availability of a complementing gene, availability of an alternative pathway or acquisition and loss of entire modules of the molecular machinery). It is noteworthy that, despite the high significance of the observed correlations between the long-term (as captured in a gene's phyletic patterns) and short-term (determined in actual knockout experiments) propensities for gene loss, the actual dependence is relatively weak. The strength of correlation between nominal variables can be represented in terms of mutual entropy, i.e., the amount of information that can be gained when the data on gene phyletic pattern is added to the data on gene knockout effect (Appendix 1). For *C. elegans* and *S. cerevisiae*, the relative information gain, i.e., the improvement in the prediction of the outcome of the genome-wide gene knockout experiment, from using phyletic patterns was calculated to be ~10.5% and ~15%, respectively. It seems notable that the link between the phyletic pattern (which represents the history of gene losses across the eukaryotic crown group) and the knockout effect are 1.5 times stronger for yeast than for the nematode; this probably reflects the greater complexity and the associated partial redundancy of the metazoan cellular machinery.

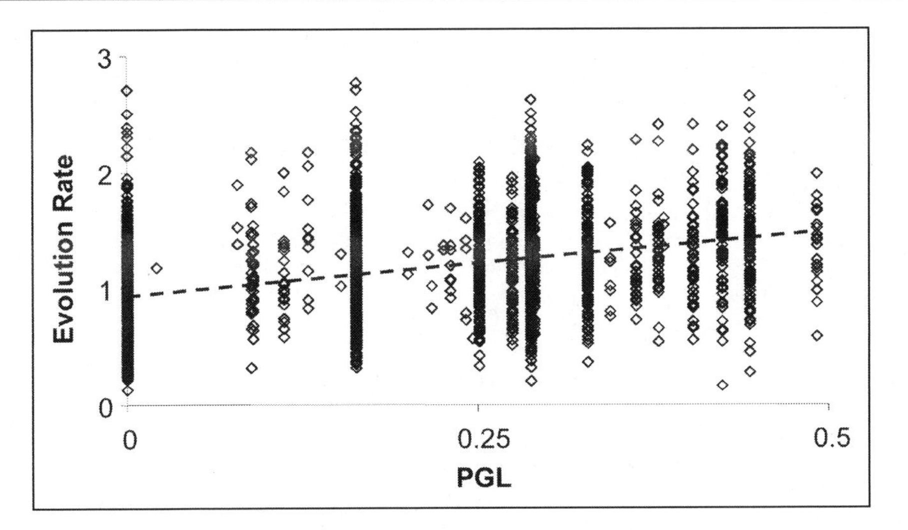

Figure 2. Correlation between sequence evolution rate and gene loss. The data was taken from reference 15. Horizontal axis—propensity for gene loss; vertical axis—evolution rate (distance from Arabidopsis protein to fungal or metazoan ortholog). Spearman rank correlation coefficient $R = 0.40$ (significant at $p \ll 10^{-10}$).

Gene Duplications and Evolution Rate

The crucial role gene duplications play in evolution had been recognized since the early days of modern evolutionary biology[21,22] and was molded into a coherent concept by Susumu Ohno in the classical 1970 book "Evolution by Gene Duplication."[23] A largely underappreciated aspect in the relationships between gene duplication and evolution of function is that several evolutionary forces, acting on a freshly duplicated pair of genes, seem to work in opposite directions. Duplication creates functional redundancy, which results in an immediate decrease of the purifying selection pressure. However, with the frequency of deleterious mutations being much higher than that of advantageous mutations, the loss of selective pressure leads to rapid "pseudogenization", making the Ohno-style neofunctionalization[23] an unlikely event. Several theoretical explanations have been proposed to resolve this apparent paradox, in particular, the subfunctionalization model, whereby young duplicate genes undergo partial loss of function, leading first to retention of both copies necessary for genetic complementation between them and, later, to functional divergence;[24] dosage effect, postulating direct selective advantage of the increase of gene (product) dosage brought about by duplication,[25] and tissue- or development stage-specific epigenetic silencing of one of the duplicates, which exposes both copies to purifying selection.[26] The observed reduction of species-wide sequence polymorphism in recently duplicated genes in Arabidopsis suggests a role of selection sweeps in initial fixation of duplications.[27] Interestingly, as a counter-point to the common notion of the creative role of gene duplication, a gene loss that "compressed" functions of two paralogs into a single copy has been suggested as the main event that "unlocked" the evolutionary path to flowering plants.[28]

Regardless of the exact nature of the relationships between gene duplication and evolution mode and rate, complex dependencies are seen in quantitative comparisons. The initial increase of evolution rate (but apparently not to the level of the neutral expectation) has been widely observed[25,29-31] although reports differ on whether the two duplicated copies typically evolve at similar[25] or significantly different[32,33] rates. Apparently, the asymmetry in the evolutionary fates of the duplicated copies extends to the patterns of expression and protein-protein interactions, and the response to environmental stress and gene disruption.[34] Large-scale studies indicate, however, that genes which have close paralogs, on average, evolve slower than singletons;[31,35] this probably reflects the stronger tendency of slower-evolving essential genes to retain a duplication for an extended period of time. Interestingly, duplication itself tends to diminish experimentally detectable fitness effect of gene disruption due to the very reason of introducing redundancy into the genetic makeup of the organism.[20,36]

Interactions between Three and More Parameters: More Than the Sum of the Parts?

Considering more than two parameters gives an additional insight into the quantitative-genomic relationships. For example, there appears to be a weak but detectable negative correlation between the evolution rate and experimentally determined number of protein-protein interactions.[5,6] Both of these parameters are correlated with expression level—highly expressed proteins tend to evolve slower and have more interactions. Accounting for the expression level brings the (already weak) correlation between the evolutionary rate and protein interactions below the significance level.[9] There is a convincing argument that the experimental detection of protein-protein interactions is strongly affected by the protein abundance; thus, the interaction data set is biased towards having an artificially high number of interaction partners for highly expressed proteins. This suggests that there might be no direct connection between the position of the protein in the interaction network and its rate of evolution. The debate that followed[7-10,37] failed, so far, to provide a definitive answer beyond the general agreement that "the large-scale data sets remain woefully noisy and incomplete."[8]

Rocha and Danchin applied multiple regression and partial correlation analysis to the data on evolution rate, expression level, functional category, essentiality and metabolic cost of genes in two model bacteria, *Bacillus subtilis* and *Escherichia coli*.[38] They showed that an indirect measure of expression level, the Codon Adaptation Index (CAI), is responsible for the major part (91-94%) of the variance in the evolution rate of bacterial genes, which is explained by multiple linear regression. Rocha and Danchin argue that, when controlled for CAI contribution, the other factors play "minor (if any) role" in determining the evolution rate of bacterial genes and explain the correlations reported by other researchers[18,19] by indirect influence of differences in expression level. While their analysis is very similar in spirit to that of Bloom and Adami,[9,10] there seems to be an important distinction: Bloom and Adami invoke an experimental bias as an explanation of the observed connection between the number of protein-protein interactions and protein abundance which indirectly explains the apparent correlation between the number of interactions and evolution rate; by contrast, Rocha and Danchin consider real correlations between three (more or less) independent variables. The former case, if solid, seems to warrant the dismissal of the observed correlations as artificial; the latter calls for development of a conceptual model taking into account the full complexity of multi-dimensional, inter-correlated data.

The "Social Status" Model

We would like to propose an idealized model that might help in developing biologically relevant null hypotheses for observed connections between quantitative measures of genome evolution and function. Firstly, let us note that it seems useful to make a distinction between "phenotypic" and "evolutionary" variables. The former, e.g., gene expression level or viability of a knock-out mutant, reflect the biology of extant organisms; by contrast, the latter, e.g., sequence evolution rate or propensity for gene loss, reflect various aspects of genome conservation and change over the course of evolution. The relationship between the phenotypic and evolutionary variables appears to have a distinct polarity: the former affect the latter because natural selection constraints or drives the evolutionary change by "testing" the organism's phenotype for fitness, but not vice versa. Phenotypic parameters directly interact with each other (e.g., codon bias of a gene affects its expression level) whereas evolutionary parameters are indirectly correlated (e.g., in accord with the neutral theory of evolution,[39] the evolutionary rates of orthologs in different lineages tend to be similar inasmuch as they perform similar functions in the respective organisms). In a sense, the phenotypic parameters provide the "input" and the evolutionary parameters represent the "output" of a biological system. Evolutionary parameters are readily produced by comparative genomic techniques (although systematic error may creep in, e.g., in calculations of evolutionary rates over long time spans) whereas most phenotypic parameters can be obtained, on genome scale, only through costly and, at this stage, highly error-prone large-scale experiments.

We suggest that the majority of the relationships between the input parameters (and, indirectly, between the output parameters) can be described with a single, composite variable which reflects the role of the gene in the cell physiology and its mode of evolution. This variable is akin to the gene's "social status in the genomic community" and relates to the importance of its functions in the overall scheme of things. "High-status" genes, which are involved in key house-keeping processes, are more likely to be higher and broader expressed, to have more interaction partners, and to produce lethal or severely impaired knockout mutants. These genes also tend to evolve slower and are less prone to gene loss across various phylogenetic lineages. "Low-status" genes are expected to be weakly expressed, have fewer interaction partners, and exhibit narrower (and less coherent) phyletic distribution. They also, on average, evolve faster and are more often lost during evolution than high-status genes.

Parameters that contribute to the status with the same sign are expected to show positive correlation between each other, whereas those that contribute in the opposite direction are expected to be negatively correlated. Thus, input parameters, which all make a positive contribution to the status (high-status genes are, generally, highly expressed, their products interact with many other proteins, their knockouts have severe fitness effects etc), are positively correlated with each other but negatively correlated with output parameters (the fast-evolving genes typically have a low status). The notion of gene status may provide a useful generator of null hypotheses regarding the connections between variables associated with functioning and evolution of genes (Fig. 3). Any deviation from the expected pattern calls for attention—to the quality of the data, the nature of the analyzed relationship, or both.

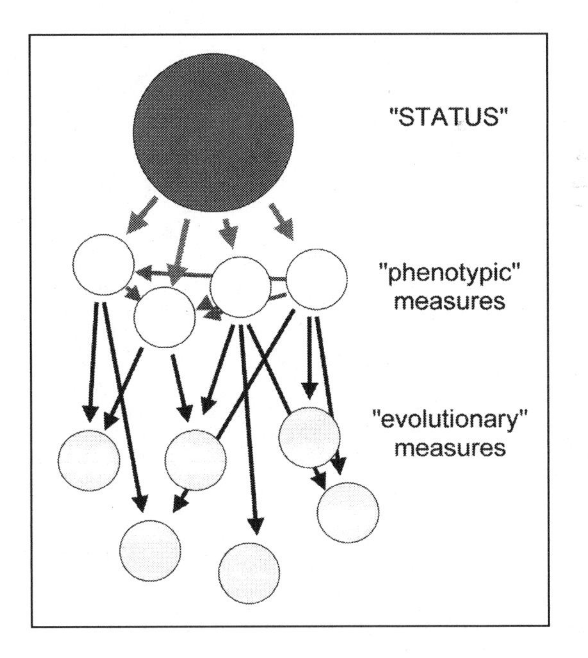

Figure 3. The "gene social status" model. Blue arrows—direct interactions between "phenotypic" (input) variables; black arrows—influence of "phenotypic" (input) variables on "evolutionary" (output) variables; red arrows—manifestation of "social status" in the "phenotypic" variables. A color version of this figure is available online at http://www.Eurekah.com.

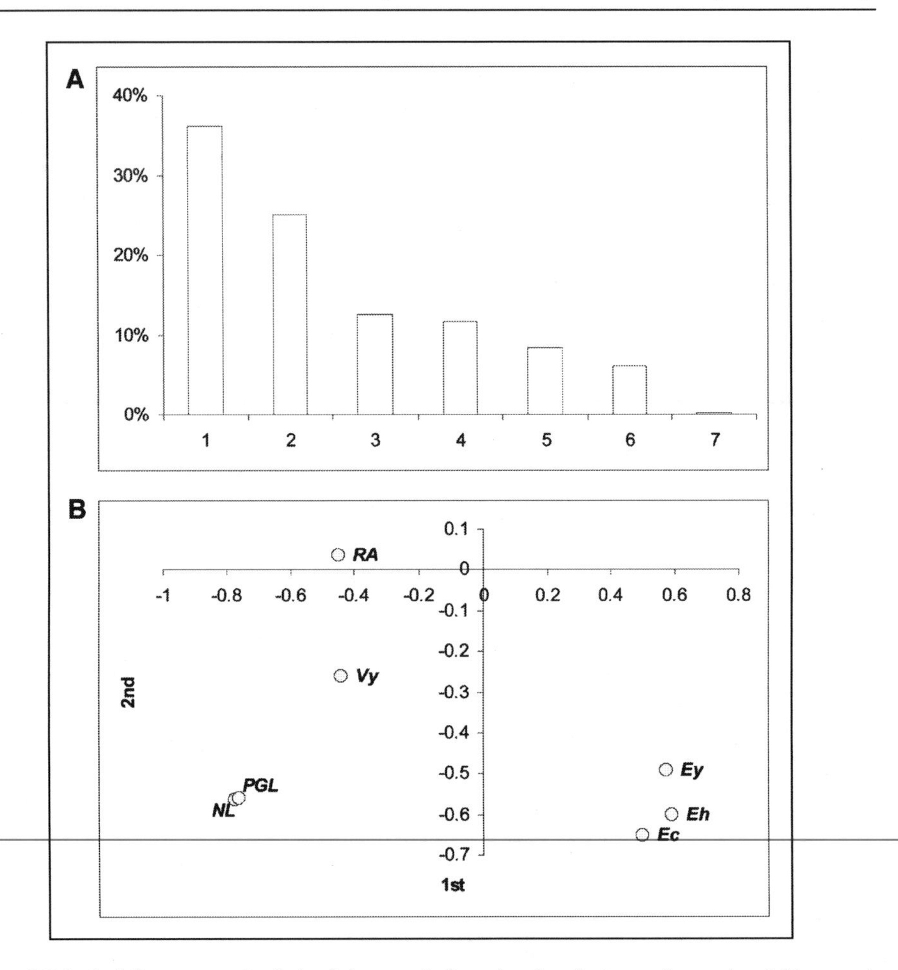

Figure 4. Principal Component Analysis of phenotypic (input) and evolutionary (output) variables associated with eukaryotic KOGs. The values of 7 variables were obtained as previously described:[15] (i) evolution rate (measured as distance from an Arabidopsis protein to fungal or metazoan ortholog), (ii) number of losses in the KOG history (reconstructed using Dollo parsimony), (iii) propensity for gene loss (essentially, number of losses, normalized for the lengths of the corresponding branches), (iv) expression level in yeast, (v) expression level in *C. elegans*, (vi) expression level in humans, and (vii) viability of yeast knock-out mutant, A) Distribution of variance among principal components. Horizontal axis—principal components; vertical axis—fraction of total variance. B) Loadings plot for the original variables in the plane of the first two principal components. Note the positive contribution of the expression level (associated with high status) and the negative contribution of evolution rate, gene loss, and viability of gene disruption mutants (associated with low status) to the first principal component. The data was from references 15 and 16.

Multi-Dimensional Structure of Expression, Evolution Rate, and Gene Loss Data

We investigated the multi-dimensional structure of expression, evolution rate, and gene loss data for a set of orthologous gene families and the correlations of these parameters with the viability of yeast knockout mutants. The principal component analysis (PCA) shows (Fig. 4) that the major direction of the data scatter, which accounts for nearly 40% of the entire variance, is formed by positive contribution from various measures of expression level (EST data for human genes and microarray data for yeast and worm orthologs) and the negative contribution from different measures of evolutionary rate (evolutionary distances between Arabidopsis proteins and their fungal or animal orthologs) and gene loss (propensity for gene loss calculated as previously described[15] or, simply, the number of losses). A coordinate on such an axis can be directly interpreted as a measure of the gene's "social status"; the fact that it is the most significant direction in terms of the data variance indicates that the "status" defined in this fashion is, indeed, important in determining the place of a gene in the data space.

Importantly, despite the fact that all pairwise correlations between the parameters are highly significant and follow the predictions of the "status" model in their signs, the overall level of interdependence between the parameters is quite low. We used the above data on the gene expression, evolutionary rate,and gene loss to predict an outcome of a gene knockout experiment in yeast (Appendix 2). As expected, slowly evolving, evolutionarily stable,and highly expressed genes are more likely to produce a nonviable phenotype compared to the genes from the opposite side of the "status" spectrum. However, the contribution of all these factors is remarkably low—using this information, the Bayesian Linear Discriminator removed only 0.5-5% from the original entropy of the gene knockout data.

Conclusions

The opportunity to analyze, systematically and quantitatively, the connections between numerous measures of genome evolution and function is one of the most alluring avenues of study opened up by the development of genomics and systems biology. Under an optimistic scenario, this might be the key to the main point of entire systems biology enterprise, transforming biology "from stamp collection to physics". Yet, it seems that any researcher who attempts to examine and evaluate the wealth of literature that has accumulated in this area in the last few years hardly can avoid a feeling of uneasiness. There seem to be too many contradicting reports on the same issue and too many high claims based on rather weak (even if statistically significant) evidence. One can easily think of at least four, certainly not exclusive, causes of this situation: (i) lack of a general conceptual framework for analysis of connections between genomic variables, (ii) the low and nonuniform quality of many types of data, (iii) inadequacy of the presently analyzed variables for understanding the connections between evolution and phenotype (we are barking on a wrong tree), (iv) the current parameters are, more or less, the best that can be measured, but they are intrinsically of limited importance for understanding those connections, which simply cannot be adequately captured by quantitative analysis (we are barking on the right tree but it is a small one).

Here we made a preliminary attempt to address problem (i) by introducing the notion of the "social status" of a gene and the distinction between "input" and "output" parameters. These are simplistic attempts on a synthesis of the information on genome-evolution-phenotype connections but they seem to work in the sense that the status concept gives unequivocal predictions on the nature of the connection (negative or positive correlation) between any two variables, and these predictions hold for the great majority of the available trials. Thus, any deviations can be construed as a signal of alarm and/or interest.

The apparent utility of the "status" concept is the flip side of the coin. The flop side comes up when we determine how much all the available information on the values of input and output parameters can improve the prediction of the outcome of a genome-scale gene knockout experiment. The improvement that could be achieved with the best possible combination of these parameters was almost shockingly small. This suggests that some combination of factors (ii)-(iv) defines the situation. The problem with the data (ii) surely is transient; there is no doubt that, within the next few years, we will witness a dramatic improvement in the completeness and accuracy of genome-wide measurements of expression, protein interactivity, and other input parameters. There is a chance that this dramatically improves the predictive power of the "gene's social status". If not, the choice will be between (iii) and (iv). The latter possibility, while perhaps discouraging, is not at all unimaginable: the principal determinants of the output values (e.g., evolutionary rate) may well lie in the features of gene and protein structure and function that cannot be captured in simple, numerical values.

***Table A1. Connection between gene knockout data and phyletic patterns for* C. elegans *and* S. cerevisiae**

	C. elegans	*S. cerevisiae*
Total entropy, bit	0.5554	0.8437
Mutual entropy, bit	0.0585	0.1252
Relative gain	10.52%	14.84%

Appendix 1. Mutual Entropy of Gene Knockout Data and Phyletic Patterns

Let p_L be the fraction of genes that produce lethal knockout mutants (obviously, there is a fraction of $1-p_L$ genes producing a viable mutant phenotype). Taking P_L and $1-P_L$ as estimates of the probability of a gene to be lethal or nonlethal, respectively; then, the total entropy that can be associated with gene knockout data is

$$H_0 = -p_L \log_2(p_L) - (1-p_L)\log_2(1-p_L)$$

Now, let us group the genes according to their phyletic patterns, and let f_i be the frequency of the i-th pattern. Let us denote the fraction of genes with lethal knockouts in the i-th pattern by p^i_L. If we think of the knockout lethality of a gene as one random variable and of its phyletic pattern as a second random variable, we can compute the conditional entropy of knockout lethality given the phyletic pattern from

$$H_1 = \Sigma f_i[-p^i_L \log_2(p^i_L) - (1-p^i_L)\log_2(1-p^i_L)]$$

The mutual entropy between these two random variables is defined as H_0-H_1; this is an accepted measure for the amount of information that each random variable carries about the other.[40] Here, we shall use the relative gain, which is a normalized version of the mutual entropy, defined as $(H_0-H_1)/H_0$.

The data on viability of gene knockout mutants were obtained from reference 41 for *C. elegans* and from reference 42 for *S. cerevisiae*. Phyletic patterns for KOGs were taken from the eukaryotic KOG database.[43]

Appendix 2. Expression Level, Evolution Rate, and Gene Loss as Predictors of Viability of Gene Knockout Mutants

We attempt to predict the viability of gene knockout mutants[41,42] using the data on expression level, evolution rate and gene loss.[15] We employed Bayesian Linear Discriminant Analysis[44] to find an optimal linear discriminant function. In brief, we compute a linear function $g(X)$, where X is a vector of variables (namely, expression level in yeast, nematode, and human, minimum and average evolutionary distance from Arabidopsis to fungi and metazoan, PGL, and number of gene losses in a KOG). For a given X, the gene knockout is predicted to be lethal if $g(X) > 0$ and nonlethal if $g(X) < 0$. The function $g(X)$ is the linear function that guarantees minimum classification error on the training dataset.

As with associating the mutant phenotype with phyletic patterns, we define the entropy of the gene knockout data as

$$H_0 = -p_L \log_2(p_L) - (1-p_L)\log_2(1-p_L)$$

where p_L is the total fraction of lethal mutants. With the prediction, obtained using Bayesian Linear Discriminator, let us define the fraction of predicted lethals as f^L, fraction of lethal phenotypes observed among predicted lethals as p^L_L (true positives), and fraction of lethal

***Table A2. Prediction of gene knockout phenotype from expression level, evolution rate and gene loss for* C. elegans *and* S. cerevisiae**

	C. elegans	*S. cerevisiae*
Total initial entropy, bit	0.7635	0.9125
Mutual entropy, bit	0.0034	0.0484
Relative information gain	0.0044	0.0530

phenotypes observed among predicted nonlethals as $p^N{}_L$ (false positives). The entropy, given the prediction, is

$$H_1 = f^L[-p^L{}_L\log_2(p^L{}_L)-(1-p^L{}_L)\log_2(1-p^L{}_L)]+(1-f^L)[-p^N{}_L\log_2(p^N{}_L)-(1-p^N{}_L)\log_2(1-p^N{}_L)]$$

Again, the mutual entropy is defined as (H_0-H_1) and the relative gain is $(H_0-H_1)/H_0$.[40]

References

1. Steinmetz LM, Davis RW. High-density arrays and insights into genome function. Biotechnol Genet Eng Rev 2000; 17:109-146.
2. Steinmetz LM, Davis RW. Maximizing the potential of functional genomics. Nat Rev Genet 2004; 5(3):190-201.
3. Hurst LD, Pal C, Lercher MJ. The evolutionary dynamics of eukaryotic gene order. Nat Rev Genet 2004; 5(4):299-310.
4. Wolfe KH, Li WH. Molecular evolution meets the genomics revolution. Nat Genet 2003; 33(Suppl):255-265.
5. Fraser HB, Hirsh AE, Steinmetz LM et al. Evolutionary rate in the protein interaction network. Science 2002; 296(5568):750-752.
6. Jordan IK, Wolf YI, Koonin EV. No simple dependence between protein evolution rate and the number of protein-protein interactions: Only the most prolific interactors tend to evolve slowly. BMC Evol Biol 2003; 3(1):1.
7. Fraser HB, Wall DP, Hirsh AE. A simple dependence between protein evolution rate and the number of protein-protein interactions. BMC Evol Biol 2003; 3(1):11.
8. Fraser HB, Hirsh AE. Evolutionary rate depends on number of protein-protein interactions independently of gene expression level. BMC Evol Biol 2004; 4(1):13.
9. Bloom JD, Adami C. Apparent dependence of protein evolutionary rate on number of interactions is linked to biases in protein-protein interactions data sets. BMC Evol Biol 2003; 3(1):21.
10. Bloom JD, Adami C. Evolutionary rate depends on number of protein-protein interactions independently of gene expression level: Response. BMC Evol Biol 2004; 4(1):14.
11. Duret L, Mouchiroud D. Determinants of substitution rates in mammalian genes: Expression pattern affects selection intensity but not mutation rate. Mol Biol Evol 2000; 17(1):68-74.
12. Pal C, Papp B, Hurst LD. Highly expressed genes in yeast evolve slowly. Genetics 2001; 158(2):927-931.
13. Zhang P, Gu Z, Li WH. Different evolutionary patterns between young duplicate genes in the human genome. Genome Biol 2003; 4(9):R56.
14. Zhang L, Li WH. Mammalian housekeeping genes evolve more slowly than tissue-specific genes. Mol Biol Evol 2004; 21(2):236-239.
15. Krylov DM, Wolf YI, Rogozin IB et al. Gene loss, protein sequence divergence, gene dispensability, expression level, and interactivity are correlated in eukaryotic evolution. Genome Res 2003; 13(10):2229-2235.
16. Jordan IK, Marino-Ramirez L, Wolf YI et al. Conservation and coevolution in the scale-free human gene coexpression network. Mol Biol Evol 2004; 21(11):2058-2070.
17. Novichkov PS, Omelchenko MV, Gelfand MS et al. Genome-wide molecular clock and horizontal gene transfer in bacterial evolution. J Bacteriol 2004; 186(19):6575-6585.
18. Hirsh AE, Fraser HB. Protein dispensability and rate of evolution. Nature 2001; 411(6841):1046-1049.
19. Jordan IK, Rogozin IB, Wolf YI et al. Essential genes are more evolutionarily conserved than are nonessential genes in bacteria. Genome Res 2002; 12(6):962-968.

20. Yang J, Gu Z, Li WH. Rate of protein evolution versus fitness effect of gene deletion. Mol Biol Evol 2003; 20(5):772-774.
21. Fisher RA. The possible modification of the response of the wild type to recurrent mutations. Am Nat 1928; 62:115-126.
22. Haldane JBS. The part played by recurrent mutation in evolution. Am Nat 1933; 67:5-19.
23. Ohno S. Evolution by gene duplication. Berlin-Heidelberg-New York: Springer-Verlag, 1970.
24. Lynch M, Force A. The probability of duplicate gene preservation by subfunctionalization. Genetics 2000; 154(1):459-473.
25. Kondrashov FA, Rogozin IB, Wolf YI et al. Selection in the evolution of gene duplications. Genome Biol 2002; 3(2):RESEARCH0008.
26. Rodin SN, Riggs AD. Epigenetic silencing may aid evolution by gene duplication. J Mol Evol 2003; 56(6):718-729.
27. Moore RC, Purugganan MD. The early stages of duplicate gene evolution. Proc Natl Acad Sci USA 2003; 100(26):15682-15687.
28. Albert VA, Oppenheimer DG, Lindqvist C. Pleiotropy, redundancy and the evolution of flowers. Trends Plant Sci 2002; 7(7):297-301.
29. Lynch M, Conery JS. The evolutionary fate and consequences of duplicate genes. Science 2000; 290(5494):1151-1155.
30. Nembaware V, Crum K, Kelso J et al. Impact of the presence of paralogs on sequence divergence in a set of mouse-human orthologs. Genome Res 2002; 12(9):1370-1376.
31. Jordan IK, Wolf YI, Koonin EV. Duplicated genes evolve slower than singletons despite the initial rate increase. BMC Evol Biol 2004; 4(1):22.
32. Conant GC, Wagner A. Asymmetric sequence divergence of duplicate genes. Genome Res 2003; 13(9):2052-2058.
33. Kellis M, Birren BW, Lander ES. Proof and evolutionary analysis of ancient genome duplication in the yeast Saccharomyces cerevisiae. Nature 2004; 428(6983):617-624.
34. Wagner A. Asymmetric functional divergence of duplicate genes in yeast. Mol Biol Evol 2002; 19(10):1760-1768.
35. Davis JC, Petrov DA. Preferential duplication of conserved proteins in eukaryotic genomes. PLoS Biol 2004; 2(3):E55.
36. Gu Z, Steinmetz LM, Gu X et al. Role of duplicate genes in genetic robustness against null mutations. Nature 2003; 421(6918):63-66.
37. Hahn MW, Conant GC, Wagner A. Molecular evolution in large genetic networks: Does connectivity equal constraint? J Mol Evol 2004; 58(2):203-211.
38. Rocha EP, Danchin A. An analysis of determinants of amino acids substitution rates in bacterial proteins. Mol Biol Evol 2004; 21(1):108-116.
39. Kimura M. The neutral theory of molecular evolution. Cambridge: Cambridge University Press, 1983.
40. Cover TM, Thomas JA. Elements of information theory. Boston: Wiley-Interscience, 1991.
41. Kamath RS, Fraser AG, Dong Y et al. Systematic functional analysis of the Caenorhabditis elegans genome using RNAi. Nature 2003; 421(6920):231-237.
42. Giaever G, Chu AM, Ni L et al. Functional profiling of the Saccharomyces cerevisiae genome. Nature 2002; 418(6896):387-391.
43. Koonin EV, Fedorova ND, Jackson JD et al. A comprehensive evolutionary classification of proteins encoded in complete eukaryotic genomes. Genome Biol 2004; 5(2):R7.
44. Duda RO, Hart PE, Stork DG. Pattern classification. Boston: Wiley-Interscience, 2000.

Index

I

L

M

N

O

P

Q

R

S

T

V

Y